DNA Sequencing: Optimizing the Process and Analysis

Jones and Bartlett Titles in Biological Science

AIDS: The Biological Basis, Third Edition
Alcamo

AIDS: Science and Society, Fourth Edition
Fan/Conner/Villarreal

Alcamo's Fundamentals of Microbiology,
Seventh Edition
Pommerville

*Alcamo's Laboratory Fundamentals of
Microbiology*, Seventh Edition
Pommerville

Aquatic Entomology
McCafferty/Provonsha

Biology: Investigating Life on Earth, Second
Edition
Avila

The Cancer Book
Cooper

Cell Biology: Organelle Structure and Function
Sadava

Creative Evolution?!
Campbell/Schopf

*Defending Evolution: A Guide to the Evolution/
Creation Controversy*
Alters

*Diversity of Life: The Illustrated Guide to the
Five Kingdoms*
Margulis/Schwartz/Dolan

Early Life: Evolution on the Precambrian Earth,
Second Edition
Margulis/Dolan

Electron Microscopy, Second Edition
Bozzola/Russell

Elements of Human Cancer
Cooper

Encounters in Microbiology
Alcamo

Essential Genetics: A Genomic Perspective,
Third Edition
Hartl/Jones

Essentials of Molecular Biology, Fourth Edition
Malacinski

Evolution, Third Edition
Strickberger

Experimental Techniques in Bacterial Genetics
Maloy

*Exploring the Way Life Works: The Science of
Biology*
Hoagland/Dodson/Hauck

Genetics: Analysis of Genes and Genomes,
Sixth Edition
Hartl/Jones

Genetics of Populations, Second Edition
Hedrick

Genomic and Molecular Neuro-Oncology
Zhang/Fuller

Grant Application Writer's Handbook
Reif-Lehrer

How Cancer Works
Sompayrac

How the Human Genome Works
McConkey

How Pathogenic Viruses Work
Sompayrac

*Human Embryonic Stem Cells: An Introduction
to the Science and Therapeutic Potential*
Kiessling/Anderson

Human Genetics: The Molecular Revolution
McConkey

The Illustrated Glossary of Protoctista
Margulis/McKhann/Olendzenski

Introduction to the Biology of Marine Life,
Eighth Edition
Sumich/Morrissey

*Laboratory and Field Investigations in Marine
Life*, Seventh Edition Reissue
Sumich/Dudley

Laboratory Research Notebooks
Jones and Bartlett Publishers

Major Events in the History of Life
Schopf

Microbes and Society
Alcamo

Microbial Genetics, Second Edition
Maloy/Cronan/Freifelder

*Missing Links: Evolutionary Concepts and
Transitions Through Time*
Martin

Oncogenes, Second Edition
Cooper

*100 Years Exploring Life, 1888–1988, The
Marine Biological Laboratory at Woods Hole*
Maienschein

Origin and Evolution Of Intelligence
Schopf

Plant Cell Biology: Structure and Function
Gunning/Steer

Plants, Genes, and Crop Biotechnology, Second
Edition
Chrispeels/Sadava

Population Biology
Hedrick

Protein Microarrays
Schena

*Statistics: Concepts and Applications for
Science*
LeBlanc

DNA Sequencing: Optimizing the Process and Analysis

Edited By
Jan Kieleczawa, Ph.D.
Wyeth Research, Cambridge, Massachusetts

JONES AND BARTLETT PUBLISHERS
Sudbury, Massachusetts
BOSTON   TORONTO   LONDON   SINGAPORE

World Headquarters
Jones and Bartlett Publishers
40 Tall Pine Drive
Sudbury, MA 01776
978-443-5000
info@jbpub.com
www.jbpub.com

Jones and Bartlett Publishers Canada
2406 Nikanna Road
Mississauga, ON L5C 2W6
CANADA

Jones and Bartlett Publishers International
Barb House, Barb Mews
London W6 7PA
UK

Copyright © 2005 by Jones and Bartlett Publishers, Inc.

All rights reserved. No part of the material protected by this copyright may be reproduced or utilized in any form, electronic or mechanical, including photocopying, recording, or by any information storage and retrieval system, without written permission from the copyright owner.

Production Credits
Chief Executive Officer: Clayton Jones
Chief Operating Officer: Don W. Jones, Jr.
President, Higher Education and Professional Publishing: Robert W. Holland, Jr.
V.P., Design and Production: Anne Spencer
V.P., Sales and Marketing: William Kane
V.P., Manufacturing and Inventory Control: Therese Bräuer
Executive Editor, Science: Stephen L. Weaver
Managing Editor, Science: Dean W. DeChambeau
Associate Editor, Science: Rebecca Seastrong
Senior Production Editor: Louis C. Bruno, Jr.
Marketing Manager: Matthew Bennett
Marketing Associate: Laura M. Kavigian
Text Design: Louis C. Bruno, Jr.
Cover Design: Anne Spencer
Composition: SNP Best-Set Typesetter, Ltd.
Printing and Binding: Malloy
Cover Printing: Malloy

Library of Congress Cataloging-in-Publication Data
DNA Sequencing: Optimizing the Process and Analysis / Jan Kieleczawa
 p.cm.
 Includes bibliographical references and index.
 ISBN 0-7637-4782-3
 1. Nucleotide sequence—Methodology. I. Kieleczawa, Jan.
 QP624.O68 2005
 572.8'633—dc22   2004048290

Printed in the United States of America
08 07 06 05 04   10 9 8 7 6 5 4 3 2 1

Brief Table of Contents

Foreword   xvii
Elaine R. Mardis

Chapter 1. Controlled Heat-Denaturation of DNA Plasmids   1
Jan Kieleczawa

Chapter 2. Effects of Various Reagents on DNA Sequencing Process   11
Jan Kieleczawa

Chapter 3. Sequencing of Difficult DNA Templates   27
Jan Kieleczawa

Chapter 4. New DNA Sequencing Enzymes   35
Sandra L. Spurgeon and John W. Brandis

Chapter 5. Beyond pUC: Vectors for Cloning Unstable DNA   55
Ronald Godiska, Melodee Patterson, Tom Schoenfeld, and David A. Mead

Chapter 6. Recombination-Based Cloning   77
Vincent Ling

Chapter 7. Plasmid Preparation Methods for DNA Sequencing   99
Parke K. Flick

Chapter 8. Optimization of Culture Growth in 96-Deep-Well Plates   117
Jan Kieleczawa

Chapter 9. Automated DNA Scanners Used in Sequencing Laboratories   123
Jan Kieleczawa

Chapter 10. Geospiza's Finch-Server: A Complete Data Management System for DNA Sequencing   131
Sandra Porter, Joseph Slagel, and Todd Smith

Chapter 11. DNA Sequencing Database: A Flexible LIMS for DNA Sequencing Analysis   143
Donald M. Koffman and Hemchand Sookdeo

Chapter 12. Good Laboratory Practices, Good Manufacturing Procedures, and Quality Assurance in the DNA Sequencing Laboratory   157
Michele Godlevski and Thalia Taylor

Chapter 13. Future of DNA Sequencing: Towards an Affordable Genome   177
Kevin McKernan

Contents

Contributors　xii
Preface　xiii
Foreword by Elaine R. Mardis　xvii

Chapter 1. Controlled Heat-Denaturation of DNA Plasmids　1
Jan Kieleczawa

Introduction　1
Experimental Design　2
DNA Sequencing　2
Temperature-Dependence of Plasmid Denaturation　3
Buffer-Strength Dependence of Plasmid Denaturation　4
Time-Course of Heat-Denaturation of DNA Plasmids　4
　of Varying Sizes
Benefits of Heat-Denaturation　6
Conclusions　9

Chapter 2. Effects of Various Reagents on DNA Sequencing Process　11
Jan Kieleczawa

Introduction　11
Experimental Design　12
Effects of Magnesium ($MgCl_2$)　12
Effect of EDTA　12
Dilution Limits of BigDye Terminator　13
Effects of Different Salts　14
Effects of Other Reagents　15
Effects of Foreign DNA and RNA　17
Effects of Primer Concentration　18
Troubleshooting of Sequencing Reactions　18
Summary　19

Chapter 3. Sequencing of Difficult DNA Templates　27
Jan Kieleczawa

Introduction　27
Materials and Methods　29
　　Materials　29
Results　31
Discussion　32

Chapter 4. New DNA Sequencing Enzymes — 35
Sandra L. Spurgeon and John W. Brandis

Introduction 35

Development of AmpliTaq DNA Polymerase FS 37
and Other Improvements

Efforts to Find Better Dye-Labeled Terminators 38

Efforts to Find Better Enzymes 39

Sources of New Polymerases 40
- Noneubacterial Genera 40
- Other Eubacterial Species: *Thermus* and *Thermotoga* 41
- Modified Taq *Pol I* 41

Methods and Materials 43
- Enzymes 43
- Sequencing Reactions 44
- Read Length 44
- Peak Height Eveness 44

Results and Discussion 44
- Evaluation of Peak Pattern Eveness and Incorporation 44
  of Labeled Dideoxynucleotides with Different Enzymes
- Salt Tolerance 45
- Ability to Read Through Difficult Regions 46

Conclusion 48

Chapter 5. Beyond pUC: Vectors for Cloning Unstable DNA — 55
Ronald Godiska, Melodee Patterson, Tom Schoenfeld, and David A. Mead

Introduction 55

Drawbacks of pUC Vectors 56
- Vector-Driven Transcription 56
- Open Reading Frames (ORFs) 57
- Repeats 58
- Replication 59
- False Postives and False Negatives 59
- Insert-Driven Transcription 60
- High Copy Number 61
- Ampicillin Selection 61
- Plasmid Mobilization 61
- Instability of AT-Rich DNA 62

Alternative Vectors 63
- pSMART 65
- pEZSeq 67
- pSMART VC 67
- pTrueBlue 68
- pZErO 70
- pTZ, pBlueScript, pGEM 70

Conclusions 70

Chapter 6. Recombination-Based Cloning — 77
Vincent Ling

Introduction 77

Recombination-Based Cloning: Primary Considerations 77
Bacteriophage P1 Cre-Based Single loxP 80
Recombination-Echo System
Bacteriophage P1 Cre-Based Dual loxP 82
Recombination-Creator System
Lambda Phage Recombinase-Gateway System 85
The Future of Recombination Cloning 94

Chapter 7. Plasmid Preparation Methods for DNA Sequencing 99
Parke K. Flick

Introduction 99
Methods for Low Throughput Preparation of Plasmid DNA 99
Methods for High Throughput Preparation of Plasmid DNA 100
TempliPhi: A New Template Preparation Method without 101
Culturing and Purification
Comparison of TempliPhi to Methods Based on Culturing 109
Summary 111

Chapter 8. Optimization of Culture Growth in 117
96-Deep-Well Plates
Jan Kieleczawa

Introduction 117
Materials and Methods 117
Experimental Parameters Optimized 118
Conclusions 122

Chapter 9. Automated DNA Scanners Used in Sequencing 123
Laboratories
Jan Kieleczawa

Introduction 123
Brief History of Some Automated Scanners 124
DNA Sequencing Instruments—Product Overview 125
Criteria for Selecting a DNA Sequencing Instrument 125

Chapter 10. Geospiza's Finch-Server: A Complete Data 131
Management System for DNA Sequencing
Sandra Porter, Joe Slagel, and Todd Smith

Introduction 131
The Geospiza Finch-Server 132
Data Management for Core Facilities 133
    Managing Sequencing Requests 133
    Sample Tracking 133
    Security 136
    Preparation of Sample Sheets and Managing 136
    Instrument-Related Data
    Management of Chromatogram Files and Quality Control 136

    Sequencing Instrument Performance 137
  Billing 137
  Data Management for Large-Scale Sequencing 137
    DNA Sequence Assembly 138
    Maintaining Updated Sequence Databases 140
    The Finch-BLAST Manager 141
  Data Management Over the Internet: iFinch 141

Chapter 11. DNA Sequencing Database: A Flexible LIMS for DNA Sequencing Analysis 143
Donald M. Koffman and Hemchand Sookdeo

  Introduction 143
    Requesting: From Scientist to Sequencer 145
    Integrated Primer Inventory and Oligo Design 147
    Automated Primer Order Generation and Completion 149
    Tube and Plate Format Runs 151
    Reporting Results 153
    Effort Reporting Based on History of Completed Reactions 154
  Summary 155

Chapter 12. Good Laboratory Practices, Good Manufacturing Procedures, and Quality Assurance in the DNA Sequencing Laboratory 157
Michele Godlevski and Thalia Taylor

  Introduction 157
  Good Laboratory Practices, Good Manufacturing Practices 158
  Standard Operation Procedures 160
  Quality Assurrance 162
    Gel-Based System 163
    Capillary-Based System 163
  Customer Service 164
  Corrective and Preventative Action 165
  Facility Requirements 167
  Training and Personnel 168
  Reagents and Solutions 168
  Maintenance and Calibration 169
  Equipment Qualifications 169
  Process and System Validation 170
  Change Control 172
  Data Collection and Archiving 172
  Documentation and Electronic Records 173
  Inspections and Audits 174
  Costs and Other Considerations for Implementation 175
  Information Resources 175

**Chapter 13. Future of DNA Sequencing: Towards an Affordable   177
Genome**
Kevin McKernan

Introduction   177
Incremental Improvements of Existing Technologies   178
Sequencing Reaction Miniaturization and Capillary   178
  Thermal Cycling
Other Means for 200-Fold Dilutions   180
Multiplexing   182
Amplification Based Methods   183
    Parallell Pyrosequencing   183
    Polonies   184
Single Molecule Sequencing by Synthesis   187
Amplification Versus Single Molecule Detection   192
Summary   194

Index   **197**

Contributors

John W. Brandis
Applied Biosystems, Foster City, California
Chapter 4

Parke K. Flick
454 Corporation, Branford, Connecticut
Chapter 7

Ronald Godiska
Lucigen Corporation, Middleton, Wisconsin
Chapter 5

Michele Godlevski
GlaxoSmithKline, DNA Sequencing Facility, Research Triangle Park, North Carolina
Chapter 12

Jan Kieleczawa
Wyeth Research, Cambridge, Massachusetts
Chapters 1, 2, 3, 8, 9

Donald M. Koffman
DMK Concepts, Inc., Winchester, Massachusetts
Chapter 11

Vincent Ling
Compound Therapeutics, Waltham, Massachusetts
Chapter 6

Elaine R. Mardis
Washington University, St. Louis, Missouri
Foreword

David A. Mead
Lucigen Corporation, Middleton, Wisconsin
Chapter 5

Kevin McKernan
Agencourt Biosciences, Beverly, Massachusetts
Chapter 13

Melodee Patterson
Lucigen Corporation, Middleton, Wisconsin
Chapter 5

Sandra Porter
Geospiza, Inc., Seattle, Washington
Chapter 10

Tom Schoenfeld
Lucigen Corporation, Middleton, Wisconsin
Chapter 5

Joseph Slagel
Geospiza, Inc., Seattle, Washington
Chapter 10

Todd Smith
Geospiza, Inc., Seattle, Washington
Chapter 10

Hemchand Sookdeo
Wyeth Research, Cambridge, Massachusetts
Chapter 11

Sandra L. Spurgeon
Applied Biosystems, Foster City, California
Chapter 4

Thalia Taylor
GlaxoSmithKline, DNA Sequencing Facility, Research Triangle Park, North Carolina
Chapter 12

Preface

Since its introduction in 1977 (2, 4), the DNA sequencing process has become a routine and powerful laboratory technique, and over the years it has evolved significantly. In the early 1980s it took two people almost two and a half years to sequence the 40,000 kbp genome of bacteriophage T7 (J. Dunn, personal communication). Now the draft sequencing of a bacterial genome of 2 to 4 Mb takes a day and the sequencing of the entire human genome (2.9 billion base pairs) in a draft form was accomplished in under a year (1, 5). This feat was possible largely because of the biochemical and technological advances induced by the Human Genome Project initiative (3). Some of these advances are described in more detail in Chapters 4, 9, and 13.

Even with these unquestionable advances and the almost routine nature of the DNA sequencing process, a number of issues remain to be solved in order to close remaining gaps in any genome. The recent focus on non-standard approaches and techniques for sequencing difficult DNA templates was evident in a number of posters presented during the last three Genome Sequencing and Analysis Conferences in San Diego (2001), Boston (2002), and Savannah (2003). Hopefully, some of these posters will be further refined and published for general use. Chapter 3 describes methods of dealing with difficult templates.

In Chapter 1, controlled heat-denaturation of plasmids is described. In itself, the incorporation of this step into a DNA sequencing protocol is an effective way to overcome failures due not only to the nature of a specific template but also to a particular host strain, the presence of external DNA, or an insufficient amount of DNA. Chapter 2 concentrates on the effects of some widely used laboratory reagents on DNA sequencing.

The success of DNA sequencing is primarily dependent on the quality and quantity of the template. Some proven and newly emerging methods of preparing DNA templates are described in Chapter 7. Chapter 8 presents optimization of conditions for growing cell cultures in 96-well deep blocks to obtain the optimal amount of plasmid DNA. In recent years, a number of cloning vectors have been developed or modified to accommodate different needs such as larger insert sizes, cloning previously unclonable fragments, or cloning "without cutting," etc. Some of these vectors are described in two articles in Chapters 5 and 6. The massive amount of sequencing data and the need for easy access to and manipulation of these data have induced many small and medium-sized DNA sequencing facilities to incorporate some form of electronic management of their respective operations. One of the commercially available LIMS (Laboratory Information Management Systems) packages is described in Chapter 10. These LIMS packages are sold "as is" and quite often need modification to adapt to a particular lab setting. We also discuss one LIMS system (Chapter 11) developed in-house that is particularly geared towards a DNA finishing core facility. Recognizing the need for "base-perfect" DNA sequencing, in fact the Federal Drug Administration's requirements (if DNA is involved) when approving a new drug, we provide a guide on how to set up and run a GLP/GMP DNA sequencing laboratory (Chapter 12). We conclude this book with a chapter (Chapter 13) on the future of DNA sequencing, concentrating primarily on technological aspects and desirable new chemistries. The challenge of achieving a $1000 (or less) cost per genome has spurred many new initiatives and this chapter takes a snapshot of the cutting edge of DNA sequencing technologies at this point in time. Without doubt, novel technologies will be developed that will make the sequencing of any genome a routine, cost-effective benchtop exercise.

This book is intended to be a practical guide for those who routinely sequence DNA. We hope that it will help to refine and enhance your operation and serve as a starting point for your own inquiries into the intricacies of the DNA sequencing process. We also hope to incorporate the newest advances in DNA sequencing in future editions and would appreciate comments and suggestions for further improvements.

Note: This book is not a textbook, and it does not provide an exhaustive list of references at the end of each chapter. We apologize for any omissions.

Acknowledgments

The early 1990s witnessed an explosion of interdisciplinary scientific activities as a result of the DOE- and NIH-sponsored human genome program, and I was truly fortunate to be in the right spot at the right time. While developing hexamer-based sequencing methodology under the

Preface

incomparable and watchful Bill Studier, I learned the true meaning of "the devil is in the details." At the same time, I had the chance to interact closely with John Dunn whose creativity and enthusiasm continually pushed me to retry, retest, and question seemingly established and proven protocols. It is not too difficult to imagine that without the influence of these two great mentors this book would not have been possible. I am also thankful to Mike Blewitt, Barbara Lade, Jutta Paparelli, and Judy Romeo for all the scientific and technical help I received while working at Brookhaven National Laboratory.

After moving to Genetic Institute in Cambridge, MA (now Wyeth Research), I again found myself in an environment conducive to further inquiries into the range of issues related to DNA sequencing technology. At the center of my interest was, and continues to be, the development of methods to deal with various difficult DNA templates. I wish to thank the entire staff of the DNA sequencing group for their help in picking up "the slack" and for other technical help while I was busily testing. Special thanks goes to Lori Haines for her critical and editorial review of these many chapters. The management at Wyeth's Genomics Department deserves my gratitude for their help and encouragement while working on this project.

This book would not be what it is without the excellent chapters contributed by so many outstanding scientists. Thank you so much, and my apology for the evasive and not so encouraging answer "soon" all those times you asked when will the book be published.

I still remember the day when, with slight apprehension, I contacted Stephen Weaver, the science editor of this volume, at Jones & Bartlett Publishers. His instantaneous and enthusiastic encouragement gave me a great boost to continue working on the book. I am also grateful to the entire editorial and production staff at Jones & Bartlett, including Rebecca Seastrong, Anne Spencer, Lou Bruno, and Dean DeChambeau.

Finally, I wish to thank my wife Carla and my children Kasia, Alex, and Michael for their patience and understanding during these past long months when I was there but not really there.

Jan Kieleczawa
Wyeth Research
Cambridge, Massachusetts
June 2004

1. International Human Genome Sequencing Consortium. 2001. Initial sequencing and analysis of the human genome. *Nature* 409:860–921.
2. Maxam, A.M., and Gilbert, W. 1977. A new method of sequencing DNA. *Proc Natl Acad Sci USA* 74:560–564.
3. National Research Council. 1988. *Mapping and Sequencing the Human Genome*. National Academy Press. Washington, DC.

4. Sanger, F., Nicklen, S., and Coulson, A.R. 1977. DNA sequencing with chain-terminating inhibitors. *Proc Natl Acad Sci USA* 74:5463–5467.
5. Venter, J.C., Adams, M.D., Myers, E.W., et al. 2001. The sequence of human genome. *Science* 291:1304–1351.

Foreword

At present, the generation of primary DNA sequence data for determining the genome sequence of any organism would appear to be a straightforward endeavor. Furthermore, given the time over which approaches and methods for "finishing" genomic sequence (e.g., producing high-quality sequence data without gaps or ambiguities) have been devised, this process should be at a point where most, if not all, problems are solved. If both statements were true, then the process of determining, to high quality, the genome sequence of any organism would be straightforward and without challenges. This is, of course, not the case, and these issues have been (and continue to be) the subject of much research activity during the past 20 years—certainly, scientists have been trying to devise approaches to solve them almost since DNA sequencing was conceptualized. The chapters in this book represent a subset of more recent activities aimed at approaching genome-sequencing-related issues, and their topics range across the broad spectrum of methods used at every point in the genome sequencing process.

Difficult-to-sequence regions and gaps in sequence coverage are impediments to high-throughput genome sequence completion, regardless of the strategy adapted. Typically, sequence gaps occur at the subcloning stage, in which structures (hairpins or other secondary structures, high GC content, repeat content) cause the subclone to be unstable (and, therefore, unrepresented). Gaps also are due to cloning bias, an inherent, albeit poorly understood aspect that is unique to different cloning vectors and may be attributed to the foreign DNA sequence itself (1). By contrast, difficult-to-sequence regions may be cloned stably but then cause low-quality data or stops in the DNA sequence ladder because of many of the same DNA structures that cause cloning gaps. Because such regions arise for different reasons and are dealt with at different points in sequence determination, approaches to solving them fall at all points along the DNA sequencing process, from beginning to end.

There are many and varied approaches to solving difficult-to-sequence regions. Typically, such a region is found by either a dramatic drop-off ("strong stop") in sequencing peak height or by a gradual decline of sequence quality (shown by decreasing peak height and resolution with increasing noise). Strong stops almost always signify a region of high secondary structure through which the polymerase cannot synthesize. There are many reasons for a gradual decline in peak height and resolution including a mono-, di-, or trinucleotide repeat or a high GC content region. Approaches to solving these regions have included reaction additives to enhance the nucleic acid denaturation properties of the reaction (2–5), different polymerases to improve proofreading or processivity attributes (6–9), nucleotide analogs (dITP, deaza dATP/dCTP) (10–13), and alternative thermal cycling parameters (including single temperature, or "isothermal," conditions) among others. Combinations of these approaches also have been used to provide successful readthrough on difficult templates.

Gap closure poses its own unique set of approaches. Typically, a sequence gap can be claimed by the polymerase chain reaction (PCR) from a larger clone (fosmid, bacterial artificial chromosome [BAC]) or directly from the genomic DNA, using directed primers that flank the gap, followed by TA subcloning and sequencing of the resulting product. A unique approach to unstable regions that result in gaps is to claim the region from a larger clone (either by PCR or restriction enzyme digestion) and to "shatter" the resulting DNA by sonication to produce agarose-gel isolated inserts of 100 to 300 base pairs that then can be propagated stably as subclones. These "mini" libraries are then sequenced and assembled to obtain the sequence within the gap (14). High copy number vectors also enable the deletion of subclone inserts, especially when repetitive or high secondary structure regions are present. One recent method for reducing the occurrence of deletions during propagation in culture is an "inducible" vector system for BACs and fosmids (15) marketed by Epicentre Technologies (Madison, WI). Here, subclones can be propagated at low copy number but induced to high copy number by the addition of L-arabinose into the growth media, just before DNA isolation. Other vector design-based approaches to producing stable subclones are described in this volume.

The identification, analysis, and solving of problem sequencing areas is ultimately dependent on the ability to assemble and view the contributing data. This ability, in turn, is dependent on accurate sample tracking throughout the production sequencing process. The advent of large scale sequencing projects has rendered both sample tracking and the analysis of problem areas to the realm of databases and the organizational power they provide.

Foreword

DNA Sequencing: Optimizing the Process and Analysis should provide a fascinating insight into the many and varied approaches to solving problem sequences, to say nothing of the appreciation one should gain that this has been, and remains, a difficult set of problems to solve.

Elaine R. Mardis
Co-Director, Genome Sequencing Center
Washington University School of Medicine
St. Louis, Missouri

1. Chissoe, S.L., et al. 1997. Representation of cloned genomic sequences in two sequencing vectors: correlation of DNA sequence and subclone distribution. *Nucleic Acids Res* 25:2960–2966.
2. Kaspar, P., Zadrazil, S., and Fabry, M. 1989. An improved double stranded DNA sequencing method using gene 32 protein. *Nucleic Acids Res* 17:3616.
3. Kristensen, T., Vass, H., Ansorge, W., and Prydz, H. 1990. DNA dideoxy sequencing with T7 DNA polymerase: improved sequencing data by the addition of manganese chloride. *Trends Genet* 6:2–3.
4. Ranu, R.S. 1994. Relief of DNA polymerase stop(s) due to severity of secondary structure of single-stranded DNA template during DNA sequencing. *Anal Biochem* 217:158–161.
5. Choi, J.S., et al. 1999. Improved cycle sequencing of GC-rich DNA template. *Exp Mol Med* 31:20–24.
6. Ye, S.Y. and Hong, G.F. 1987. Heat-stable DNA polymerase I large fragment resolves hairpin structure in DNA sequencing. *Sci Sin [B]* 30:503–506.
7. Bechtereva, T.A., Pavlov, Y.I., Kramorov, V.I., Migunova, B., and Kiselev, O.I. 1989. DNA sequencing with thermostable Tet DNA polymerase from *Thermus thermophilus*. *Nucleic Acids Res* 17:10507.
8. McClary, J., Ye, S.Y., Hong, G.F., and Witney, F. 1991. Sequencing with the large fragment of DNA polymerase I from *Bacillus stearothermophilus*. *DNA Seq* 1:173–180.
9. Reeve, M.A. and Fuller, C.W. 1995. A novel thermostable polymerase for DNA sequencing. *Nature* 376:796–797.
10. Labeit, S., Lehrach, H., and Goody, R.S. 1986. A new method of DNA sequencing using deoxynucleoside alpha-thiotriphosphates. *DNA* 5:173–177.
11. Carlberg, C., Quaas, R., Hahn, U., and Wittig, B. 1987. Sequencing refractory GC rich regions in plasmid DNA. *Nucleic Acids Res* 15:2779.
12. Joho, K.E. DNA sequencing artifacts in dITP reactions containing gene 32 protein. 1989. *Nucleic Acids Res* 17:7111.
13. Seela, F., et al. 2001. Fluorescent DNA: the development of 7-deazapurine nucleoside triphosphates applicable for sequencing at the single molecule level. *J Biotechnol* 86:269–279.
14. McMurray, A.A., Sulston, J.E., and Quail, M.A. 1998. Short-insert libraries as a method of problem solving in genome sequencing. *Genome Res* 8:562–566.
15. Wild, J., Hradecna, Z., and Szybalski, W. 2002. Conditionally amplifiable BACs: switching from single-copy to high-copy vectors and genomic clones. *Genome Res* 12:1434–1444.

1 Controlled Heat-Denaturation of DNA Plasmids

Jan Kieleczawa
Wyeth Research, Cambridge, MA

Introduction

Unless the DNA template to be sequenced is single-stranded (ss DNA), the two strands in plasmid DNA (ds DNA) and in polymerase chain reaction (PCR) templates must be separated for the priming event and extension to occur. Prior to the introduction of thermostable DNA polymerases in the sequencing process (4, 5), the strand separation was accomplished by heat-denaturation (22°C to 85°C) of plasmids in the presence or absence of 0.1 to 0.3 N NaOH, followed by neutralization, alcohol precipitation and subsequent resuspension in the desired solution (2, 3, 8, 10). Though effective, these steps are cumbersome, time-consuming, and they are, in this author's opinion, far from optimal. In fact, most denaturation conditions described in publications (2, 3, 8, 10, and references therein) were replicated and data evaluated using agarose gel electrophoresis and DNA sequencing. In all cases, the transition from ds to ss form is only partial, if any, and in some cases additional, nonsequencable, bands are formed (data not shown). Table 1.1 shows the sequence data for four different NaOH-induced plasmid denaturation protocols and the denaturation conditions recommended in this work.

The results presented in this section are valuable to those who still use manual radioactive plasmid DNA sequencing with nonthermostable polymerases where the initial strand separation is an essential step. It also would be relatively easy to incorporate such a controlled heat-denaturation step into any process flow in high-throughput DNA sequencing centers that rely heavily on automation. As shown in Chapter 3 controlled heat denaturation is one of the most effective ways to sequence through many different categories of difficult DNA templates. In addition, it is possible to use this protocol in general PCR technology to improve the quality of the product and to further reduce the amount of the initial DNA template. Any other DNA technology that relies on effective strand separation could benefit from this protocol.

Experimental Design

All heat-denaturing experiments were carried out in a PTC-200 thermocycler (MJ Research, Waltham, MA) in 200 µl PCR tubes covered with caps. The denaturation step was performed on 200 ng of each studied DNA in a final volume of 20 µl in 10 mM Tris/Cl pH 8, 0.01 mM EDTA (TE_{sl}). During reaction preparation, sample tubes were stored on ice. All of the tubes were then placed in a preheated thermocycler block set at 98°C, unless otherwise stated. At specified time intervals, tubes were quickly withdrawn from the cycler and placed on ice until the end of series. Samples were briefly centrifuged and 2.2 µl of 10× DNA loading buffer was added and the entire sample loaded onto a 1% agarose gel (1 × TAE buffer/0.5 µg/ml etydium bromide, agarose gel size 14 × 11 cm) as described in (6). Samples were electrophoresed for approximately two hours at 100 to 150 V. Incubation in water for 30 minutes (in a cold room) removed excess etydium bromide. The intensities of the fluorescently stained bands were measured and quantified using EDAS120 Kodak 1D Image Analysis Software (Eastman Kodak Company, Rochester, NY). The time at which 75% of the ds DNA is converted to ss DNA was determined for each plasmid from their spectral characteristics. Initially, similar experiments were carried out in 0.1, 0.2, and 0.3 N NaOH. However, the need for careful neutralization and possible precipitation steps as well as the formation of additional DNA bands (mostly the collapsed form-results not shown here) rendered such an approach impractical, and it was therefore abandoned.

DNA Sequencing

Unless otherwise indicated, the DNA sequencing was carried out as follows. An aliquot (0.2 µg or as indicated in a specific experiment) of plasmid DNA was combined with 1 µl of 5 µM primer and 3 µl of twofold diluted ABI PRISM BigDye Terminator Cycle Sequencing Ready Reaction Kit mix (version 3.0). The volumes were adjusted to 10 µl with 10 mM Tris-HCl (pH 8.0) or water, and amplification reactions were performed on PTC-225 cycler (MJ Research, Waltham, MA) for 25 cycles (96°C for 10 seconds, 50°C for 5 seconds and 60°C for 4 minutes). Twenty µl of water was added to the reactions and the excess dye was removed by gel filtration on a 96-well Millipore filter plate with G-50 beads. The samples

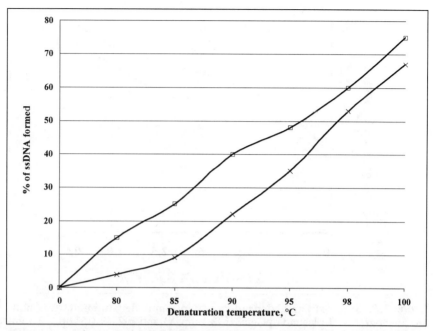

Figure 1.1. Temperature-dependence of plasmid denaturation. Two hundred nanograms of plasmids pGem3zf (3.2 kbp; ×) and pf2549 (16 kbp, □) were subjected to heat-denaturation for five minutes at indicated temperatures. All samples were then processed as described in the text.

were heat-denatured for two minutes at 90–95°C and electrophoresed on ABI3100 Genetic Analyzer (Applied Biosystems, Foster City, CA) equipped with 50- or 80-cm capillary arrays under conditions recommended by the manufacturer.

Temperature-Dependence of Plasmid Denaturation

In TE_{sl} hardly any transition from the ds to ss form occurs at temperatures below 80°C. Figure 1.1 shows an example of two plasmids denatured at various temperatures. It is apparent that only at temperatures above 90°C does any appreciable ds to ss conversion take place. For practical reasons, 98°C was selected for further experiments; however, slightly lower or higher temperatures can be used with proper time adjustments that can be estimated from Figure 1.1.

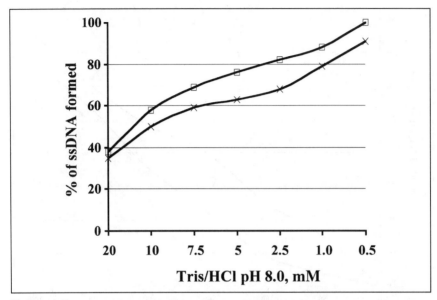

Figure 1.2. Buffer-strength dependence of plasmid denaturation. Two hundred nanograms of plasmids pGem3zf (3.2 kbp, ×) and pf2549 (16 kbp, □) were subjected to heat-denaturation for five minutes at 98°C in indicated buffers. All samples were then processed as described in the text.

Buffer-Strength Dependence of Plasmid Denaturation

The amount of salts present during heat-denaturation affects the time needed for optimal transition from ds to ss DNA. Figure 1.2 shows a denaturation experiment for two different plasmids under varying salt conditions (0.5 to 20 mM Tris/Cl buffer pH 8.0). It is worth noting that in a set period of time, almost twice as much DNA is converted from ds to ss in 5 mM Tris/Cl compared to 20 mM Tris/Cl. As a rule of thumb, the denaturation time in water or in the presence of salts below 1 mM is about half that required in 10 mM Tris. Because we recommend storing plasmid DNA in 10 mM Tris/0.01 mM EDTA (see also Chapters 7 and 9), we selected these conditions as the default for all subsequent experiments.

Time Course of Heat-Denaturation of DNA Plasmids of Varying Sizes

A series of high-quality DNA plasmids with a size range of 3 to 20 kbp was subjected to heat denaturation at 98°C as described above. Plotting

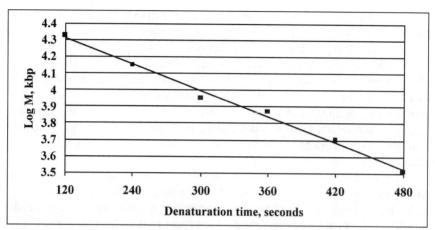

Figure 1.3. Size-dependence of plasmid denaturation. Two hundred nanograms of each of the following plasmids was denatured in a time-dependent manner: pGem3zf (3.2 kbp), pRJR1 (4.84 kbp), M13mp18 RF (7.25 kbp), p380 (9.0 kbp), FES7 (14.0 kbp), and p2000 (20.8 kbp). Similar linear data were obtained for plasmids denatured in 0.2 N NaOH and for different levels of denaturation (not shown).

the time points at which 75% of each plasmid was converted from ds to ss form vs. the log of DNA mass resulted in a linear relationship, shown in Figure 1.3. The following experimental equation was derived from this graph:

$$DT_{75\%}(\text{min}) = 7.5 - 1[(A - 3.2)/2.5]$$

where $DT_{75\%}$ = the denaturation time at which 75% of ds DNA is converted into ss form; A = size (in kbp) of the plasmid to be denatured; 3.2 is the size of pGem3zf (in kbp); and 2.5 is the factor derived from Figure 1.3. It refers to the fact that for any plasmid bigger than pGem3zf, one needs to subtract 1 minute per multiple of 2.5 kbp from 7.5 minutes to achieve a similar level of denaturation. The time at which 75% of ds DNA is converted into the ss form was selected for safety, as continuing the denaturation for longer times will lead to the degradation of the DNA.

The EDTA, at 1 mM, substantially inhibits the transition from ds to ss form and above 5 mM EDTA, no transition at all is observed. Similar inhibitory effects are seen with NaCl above 50 mM and $MgCl_2$ above 1 mM (not shown).

For plasmids of very similar sizes it is possible to precisely calculate the time needed for the desired level of denaturation. However, in a typical DNA sequencing core laboratory, one very rarely deals with such an idealized situation. Plasmids to be sequenced most often range in size from 4 to 10 kbp. It would be impractical to individually denature plasmids in each size range; therefore, we recommend using an average denaturation time of three to five minutes. If plasmids are delivered in water, one needs to halve the denaturation time compared to the time for DNA stored in TE_{sl}. However, the denaturation of plasmids in water leads to the formation of additional "non-sequence-able" DNA bands (not shown) and, although effective and used in many laboratories, it is not highly recommended based on the data presented in this work. In addition, long-term storage of plasmids in water leads to depurination (1) and may lower the sequence quality of DNA.

In general, a heat-denaturation step is not needed for PCR fragments and linear DNAs; however, a short incubation (30–60 sec) occasionally may improve the quality of the sequencing results (not shown). The denaturation data presented here relate to typical plasmids. When plasmids containing extremely high GC-rich or CTT-rich inserts were subjected to similar denaturation experiments, the time needed for effective conversion of ds to ss form was on the order of 20 to 30 minutes. We are still evaluating this unusual behavior.

Benefits of Heat-Denaturation

There are a number of benefits to heat-denaturation of plasmid DNAs:

1. Less DNA is needed to obtain optimal read lengths. During the denaturation step almost all supercoiled and nicked forms of the DNA template are converted to the sequence-able ss form. Figure 1.4 shows an example of the titration of the amount of DNA (pGem3zf) versus read length using the standard protocol and the protocol with a heat-denaturation step. DNA sequencing reactions were run on an ABI377 DNA sequencer (Applied Biosystems, Foster City, CA). For comparison, a similar titration is shown for heat-denatured samples run on an ABI3100 with a 50-cm capillary array. Note that two- to threefold less DNA is needed when a heat-denaturation step is included. For the ABI3100 a much broader usable concentration range is observed. This is especially advantageous in core laboratory settings where processed DNA samples are rarely prepared using one standard protocol and a broad range of DNA concentrations are common. Similar titration results (not shown) were obtained on a number of plasmids in the 3 to 20 kbp size range.
2. Sequencing of previously "unsequence-able" plasmids. Sometimes, inclusion of the heat-denaturation step is the only possible way to get clearly readable signal (refer to Chapter 3). We have encountered at least two scenarios:

Chapter 1. Controlled Heat-Denaturation of DNA Plasmids

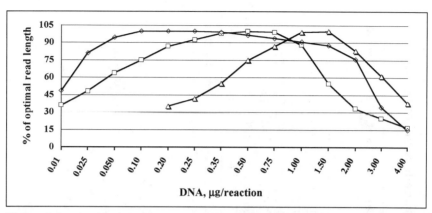

Figure 1.4. Amount of DNA versus read length: effect of heat-denaturation. Varying amounts of pGem3zf (M13 reverse primer) were sequenced using the standard (△) or modified protocol (□) and samples were run on either an ABI377 DNA sequencer or an ABI3100 (◇). For DNA samples run on an ABI3100, only samples prepared with the modified protocol are shown.

   a. No chromatogram was produced using the standard protocol and clear data when a heat-denaturation step is included.
   b. Unreadable chromatograms are obtained when using the standard protocol and clear data when heat-denaturation step is included.
3. Removing most of the secondary structures in DNA. This is the most likely reason that no chromatograms were produced in case number 1 (above), as some regions in this template are over 90% GC-rich.
4. More uniform and longer read lengths. Sixteen identical samples of pGem3zf were subjected to sequencing using standard or modified protocols. The average signal strength (in relative fluorescent units) and the read length ($Q \geq 20$) for standard protocol were 50 ± 22 and 559 ± 230, respectively. For modified protocol (5 minutes heat-denaturation at 98°C) these values were 100 ± 40 and 742 ± 13, respectively. The purification of sequencing products was done using in-house prepared G-50 sephadex columns in 96-well Millipore filter plates (Multi Screen-HV plates, Cat# MAHVN4510, Millipore, Bedford, MA). For additional data see Table 1.1.
5. Lowering the number of cycles needed for optimal results. In standard DNA sequencing protocols, 25 to 30 cycling steps are recommended to obtain optimal signal strengths and read lengths. Because of the efficient transformation to ss DNA when a heat-denaturation step is included, the number of cycles can be reduced to 10 to 15, therefore reducing by half the time needed to perform this step (data not shown).

Table 1.1. Comparison of read lengths and signal strength for varying DNA concentrations and different protocols.

Denaturation conditions → ↓ DNA (ng)		This work/H$_2$O[1]		This work/TE$_{sl}$[2]		Ref. 3	Ref. 8[3]	Ref. 8[4]	Ref. 10
		No HD	+HD	No HD	+HD				
25	Q ≥ 20	489 ± 87	744 ± 53	488 ± 84	724 ± 71	268 ± 62	284 ± 124	223 ± 92	0
	ST	13 ± 3.7	32.1 ± 6.1	12.3 ± 2.5	27.2 ± 6.0	8.8 ± 2.2	10.6 ± 3.8	8.6 ± 2.0	5.8 ± 0.5
50	Q ≥ 20	585 ± 177	764 ± 44	625 ± 106	784 ± 59	577 ± 44	409 ± 59	436 ± 92	339 ± 90
	ST	15.6 ± 5.2	47.0 ± 8.6	16.9 ± 3.4	45.8 ± 19.8	12.6 ± 2.5	10.5 ± 2.2	9.4 ± 1.8	7.6 ± 1.2
200	Q ≥ 20	837 ± 25	843 ± 22	724 ± 71	866 ± 29	792 ± 39	663 ± 113	621 ± 76	712 ± 41
	ST	59.8 ± 9.8	119.2 ± 14	41.7 ± 7.4	123.3 ± 20	65.1 ± 2.2	43.1 ± 9.7	29.8 ± 8.2	26.1 ± 5.9

Varying amounts of pGem3zf plasmid DNA (four samples for each condition) were pretreated differently before subjecting them to cycle sequencing. Processed sequencing reactions were run on ABI3100 equipped with 80cm capillary array. The data was evaluated in terms of Q ≥ 20 read length (RL) and signal strength (ST in fluorescent units).

[1] DNA was resuspended in water and sequenced using standard protocol (No HD = no heat denaturation), or samples were heat denatured (+HD) for 2.5 minutes at 98°C before cycle sequencing.

[2] DNA was resuspended in 10mM Tris/0.01 mM EDTA and sequenced using standard protocol (No HD = no heat denaturation), or samples were heat denatured (+HD) for 7.5 minutes at 98°C before cycle sequencing.

[3] DNA was denatured at room temperature (22°C) for five minutes in the presence of 0.2N NaOH. Following this treatment, equal concentration of HCl was added to neutralize this solution. The mixtures were supplemented with 40 µl of water, 6 µl of 3M sodium acetate pH 6.0, and 160 µl 95% ethanol and were centrifuged for 30 minutes at 3400× g. Supernatants were discarded, and precipitates were washed twice with 200 µl of 70% ethanol. Finally samples were dried for 15 minutes at 65°C and resuspended in 6 µl water, 1 µl of 5 µM M13 reverse primer, and 3 µl of twofold diluted BigDye terminator V3.0 and electrophoresed as described above.

[4] As above, but the denaturation was at 85°C.

The other denaturation conditions were as described in the indicated references.

Chapter 1. Controlled Heat-Denaturation of DNA Plasmids

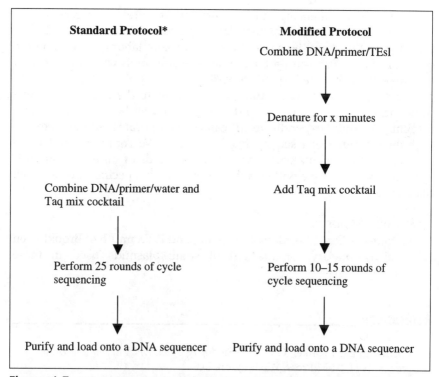

Figure 1.5. Schematic representation of standard and modified DNA sequencing protocols. ABI protocols recommend using 8 µl of Taq mix cocktail, 0.2 to 0.5 µg plasmid DNA, 1 µl of 3.2 µM primer, and water to a total volume of 20 µl. Modified protocol recommended in this work: 2 µl of Taq mix cocktail, 0.1 to 0.5 µg (default = 0.25 µg) of plasmid DNA (or 0.005–0.1 µg of PCR fragment; default = 0.05 µg), 1 µl of 5 µM primer and TE_{sl} to total volume of 10 µl. Magnesium is adjusted to a final concentration of 2 mM. In both cases we used the same cycling protocol: 10 seconds at 96°C/5 seconds at 50°C/4 minutes at 60°C. Repeat this cycle 10 to 24 times and purify sequencing reactions using ethanol precipitation or G-50 96-well sieving plate as described in Chapter 2. The number of cycling steps can be adjusted depending on the particular sequencing machine, clean up method, and the quantity of the starting DNA template used in the sequencing. A good starting point is to use the manufacturer's recommended conditions and then optimize for your particular laboratory.
*As recommended in ABI manuals.

Conclusions

The experiments described above led us to introduce an additional step into a standard ABI-like DNA sequencing protocol as shown in Figure 1.5. The adapted protocol not only reduced the amount of template needed to carry out DNA sequencing but also resulted in a greater success rate by eliminating many causes of template "difficulty." For example, before introducing this step, the success rate in our core laboratory was on the order of 65% to 70%, and after the heat-denaturation was implemented the success rate climbed to 80% to 90%.

Adding this simple denaturation step leads to the increased uniformity and longer read lengths, and it is achieved without any additional reagents or capital expenditures. It can be incorporated into any core or high-throughput DNA sequencing laboratory. We continue similar denaturation experiments with a variety of plasmids of different compositions and hope that we will be able to derive more specific rules for each distinct category.

Acknowledgments

I wish to thank Drs. C. Anderson, J. Dunn, and P. Freimuth of Brookhaven National Laboratory for the gift of some plasmids used in these experiments.

References

1. Friedberg, E.C., Walker, G.C., and Siede, W. 1995. *DNA Repair and Mutagenesis*. W.H. ASM Press. Washington, D.C.
2. Haltiner, M., Kempe, T., and Tjian, R. 1985. A novel strategy for constructing point mutations. *Nucleic Acids Res* 13: 1015–1025.
3. Hattori, M., and Sakaki, Y. 1986. Dideoxy sequencing method using denatured plasmid templates. *Anal Biochem* 152: 232–238.
4. Lee, J-S. 1991. Alternative dideoxy sequencing of double-stranded DNA by cyclic reactions using *Taq* polymerase. *DNA Cell Biol* 10: 67–73.
5. Murray, V. 1989. Improved double-stranded DNA sequencing using the linear polymerase chain reaction. *Nucleic Acids Res* 17: 8889.
6. Sambrook, J., and Russell, D.W. 2001. *Molecular Cloning*, 3rd ed. CSH Laboratory Press. Cold Spring Harbor, NY.
7. Sauer, P., Muller, M., and Kang, J. 1998. Quantitation of DNA. *Qiagen News* 2: 23–26.
8. Toneguzzo, F., Glynn, S., Levi, E., et al. 1988. Use of a chemically modified T7 DNA polymerase for manual and automated sequencing of supercoiled DNA. *BioTechniques* 6: 460–469.
9. Wilfinger. W.W., Mackey, K., and Chomczynski, P. 1997. Effect of pH and ionic strength on the spectroscopic assessment of nucleic acid purity. *BioTechniques* 22: 474–481.
10. Yie, Y., Wei, Z., and Tien, P. 1993. A simplified and reliable protocol for plasmid DNA sequencing: fast miniprep and denaturation. *Nucleic Acids Res* 21: 361.

2 Effects of Various Reagents on DNA Sequencing Process

Jan Kieleczawa
Wyeth Research, Cambridge, MA

Introduction

Despite tremendous technical advances, widespread use, and relative simplicity, DNA sequencing still requires considerable technical skills, expensive equipment, and attention to detail. A typical small academic or other core DNA sequencing laboratory may have to deal with high personnel turnover (and hence, potential for inexperience), modified or custom DNA preparation protocols, and a lack of specialized knowledge of the factors influencing the sequencing process. Quite often, the presence of these factors in DNA preparation leads to suboptimal quality of sequence data. This may be acceptable in a high-throughput, factory-like DNA sequencing centers where the overall statistics of sequence throughput is the preeminent factor. In addition, the sequencing process is often optimized for one particular set of protocols. However, in a core/academic laboratory setting, where numerous DNA preparation protocols are used, a 10% overall failure rate may translate into 100% failure for a particular individual or project.

This chapter examines the effects of popular laboratory reagents on sequencing outcome. All of these reagents can be used in DNA preparation methods, purification of PCR products, primer synthesis processes or other laboratory methods that lead towards the submission of a sample for sequencing. Some of these results can be found in published literature (3, 4, 5, and references therein) or in manuals accompanying DNA sequencing kits and protocols (1, 6).

All data presented in this chapter were independently produced by this author's laboratory using BigDye V3.0 chemistry, and similar data were obtained using the original FS and BigDye V 1.0 chemistries (not shown). At the end of this chapter, we provide a troubleshooting guide for a typical core sequencing laboratory.

DNA Sequencing: Optimizing the Process and Analysis
Edited by Jan Kieleczawa
©2005 Jones and Bartlett Publishers

Experimental Design

Unless otherwise stated, all experiments used 0.5 µg of pGem3zf plasmid (purchased from Promega, Madison, WI) and 1 µl of 5 µM M13 reverse primer. The sequencing protocol was as follows: combine the DNA, primer, the tested reagent, and the TE_{sl} (10 mM Tris, 0.01 mM EDTA, pH 8.0) in a total volume of 7 µl. Heat-denature samples for 7.5 minutes at 98°C, place them on ice, and add 3 µl of diluted BigDye terminator V 3.0 (purchased from Applied Biosystems, Foster City, CA). The final volume of the DNA sequencing reactions was 10 µl and the final dilution factor for BigDye terminator is 4×. The $MgCl_2$ was kept at a constant concentration of 2 mM, unless otherwise indicated. The cycling protocol was carried out in a PTC-200 thermal cycler (MJ Research, Waltham, MA) in 200 µl PCR thin-wall tubes using the following cycling regime: [10 seconds at 96°C/5 seconds at 50°C/4 minutes at 60°C] × 25. The sequencing reactions were purified using ethanol precipitation (3) to remove unincorporated dyes. Each data point is an average of three to six repeats in two separate experiments. The DNA sequencing was performed using an ABI377 or ABI3100 (Applied Biosystems). The read lengths are expressed in terms of 1% error rates (the longest read with a 1% error rate, which corresponds to a Q20 phred score; see also Chapter 4 for the definition of phred $Q \geq 20$). The typical $Q \geq 20$ values for controls were in the range of 550 to 600 bases. In the following graphs, inhibition/stimulation is expressed as a percentage of the control for a given experiment.

Effects of Magnesium ($MgCl_2$)

Magnesium is an essential component in stimulating activity of DNA polymerases. For polymerases used in the DNA sequencing, the optimum concentration is around 2 mM (1, 3, 4, 6).

Figure 2.1 shows the effects of magnesium ions on sequencing. The indicated amounts of $MgCl_2$ were present or absent during heat-induced strand separation (see Chapter 1 for details of heat-induced DNA denaturation). About 20% longer read lengths can be achieved when magnesium ions are absent during heat-denaturation. This is most likely due to more effective strand separation without Mg ions during the denaturation step.

Effect of EDTA

Ethylenediaminetetraacetic acid (EDTA) is a chelating agent, and it reduces the effective concentration of free magnesium ions, which are

Chapter 2. Effects of Various Reagents

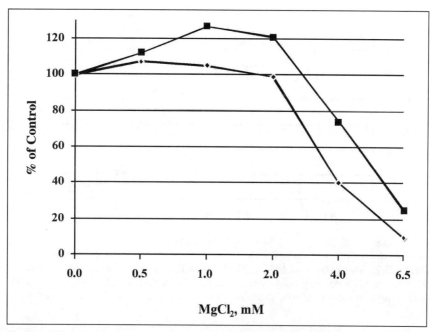

Figure 2.1. The effect of magnesium. Plasmid DNA (pGem3zf) was heat-denatured for 7.5 minutes at 98°C in the absence (■) or presence (♦) of MgCl$_2$ at indicated concentrations. Following this treatment samples were sequenced as described in the text.

necessary for full activity of DNA polymerases. Figure 2.2 shows the effect of EDTA, either absent or present, during the heat-denaturation step. When the molar ratio of EDTA to MgCl$_2$ is close to 1, almost all DNA polymerase activity is inhibited. This has practical implications, as some DNA preparation methods call for the final resuspension of DNA in TE buffer (typically 10 mM Tris-HCl pH 7–8, 1 mM EDTA). When DNA sequencing is one of possible requirements, resuspension in TE buffer should be avoided. Instead, DNA should be resuspended in a buffer with EDTA < 0.1 mM. We recommend using super low TE (TE$_{sl}$) buffer (10 mM Tris-HCl, pH 8, 0.01 mM EDTA). In addition, resuspension in water should be avoided if the DNA is going to be stored for prolonged periods of time (see also Chapter 1).

Dilution Limits of BigDye Terminator

With high-quality templates and magnesium concentration maintained at 2 mM, BigDye can be diluted approximately 10- to 15-fold (under the con-

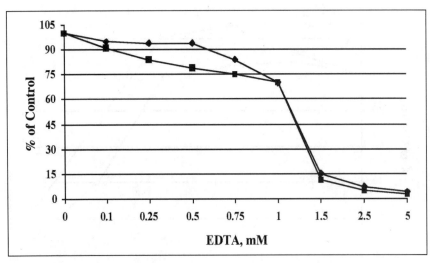

Figure 2.2. The effects of EDTA. EDTA at indicated concentrations was absent (■) or present (♦) during heat-denaturation (7.5 min at 98°C) of pGem3zf plasmid. Following this treatment, the samples were sequenced as described in the text.

ditions described above) with no apparent effect on read length. When magnesium is not close to optimal, the BigDye dilution factor should be less than fivefold. When dealing with different DNA preparations and variable template concentrations we recommend that the dilution factor of BigDye be kept at 4, otherwise sequencing may yield low quality or no data (results not shown). Figure 2.3 shows the dilution limits at constant (2 mM) and variable magnesium concentrations. The dilution factor of BigDye can be much higher than that described earlier; however, this generally requires the reduction of reaction volumes to sub-microliters, using specialized robotics, thermocyclers and high-density plates. In addition, the newest DNA sequencers (e.g., ABI 3100/ABI3730) are more sensitive compared to the ABI377, which allows for much higher dilution of the Taq mix. For optimal results, it is advisable that the dilution limits for BigDye be tested in the individual laboratory especially regarding the type of the DNA preparation methods being used.

Effects of Different Salts

Various salts are used during DNA preparation and other molecular laboratory techniques. Their carry-over or intentional presence can have a significant effect on the quality of sequence data. Figure 2.4 shows the effect of a number of salts on read length. At very low concentrations

Chapter 2. Effects of Various Reagents

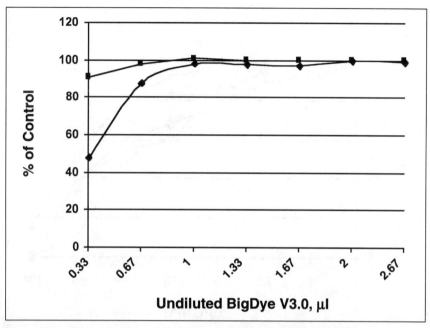

Figure 2.3. The dilution limits of dye-terminator BigDye™ V3.0. Plasmid DNA (pGem3zf) was heat-denatured for 7.5 min at 98°C and sequenced using variable (♦) or constant 2 mM $MgCl_2$ (■). The magnesium concentration for (♦) series varied from 0.25 mM to 2 mM in 0.25 mM increments.

(<2.5 mM) none of the tested salts had an appreciable effect on sequence quality and read length. Higher salt concentrations significantly decreased read lengths (20% to 50% at 25 mM) and at 100 mM the activity of DNA Taq polymerase was completely inhibited (except for ammonium acetate).

Effects of Other Reagents

A variety of other reagents can be directly or indirectly used in connection with the sequencing process. Figure 2.5 shows the effects of dimethyl sulfoxide (DMSO), alcohols, acetamide (sometimes used to PCR GC-rich templates), acetonitrile (used during oligonucleotide synthesis), and glycerol on sequencing. At concentrations of up to 5% DMSO, acetamide, acetonitrile, and glycerol have no effect on optimal read length. For some reagents (DMSO, acetamide, and glycerol) even 10% has no appreciable effect. Only at 15% is a significant decrease in read length observed. Increasing concentrations of ethanol and isopropanol gradually decrease

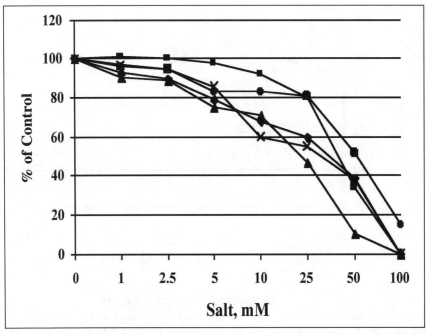

Figure 2.4. The effects of different salts. Plasmid DNA (pGem3zf) was heat-denatured for 7.5 min at 98°C in the presence of varied concentrations of different salts. Following this treatment samples were sequenced as described in the text. The salts were: ammonium acetate (●), cesium chloride (■), potassium chloride (✗), sodium acetate (♦), and sodium chloride (▲).

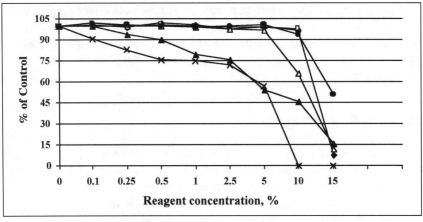

Figure 2.5. The effects of other reagents. Plasmid DNA (pGem3zf) was heat-denatured for 7.5 minutes at 98°C in the presence of varied concentrations of different reagents. Following this treatment samples were sequenced as described in the text. The reagents were: acetamide (●), acetonitrile (◇), DMSO (□), ethanol (▲), glycerol (♦), and isopropanol (✗).

Chapter 2. Effects of Various Reagents

the read length and at 5%, both of these molecules reduce the read length by half. An even stronger effect is seen with ammonia (e.g., used in the final steps of oligonucleotide synthesis): a concentration of 0.25% inhibits DNA polymerase activity by 60% and at 1% polymerase activity is completely abolished (data not shown).

Effects of Foreign DNA and RNA

Quite often "unwanted" chromosomal DNA and RNA are present in plasmid preparations. This causes errors during the determination of DNA concentration and the amount of "desired" template can be grossly underestimated, resulting in suboptimal data quality. Even if the template concentration was estimated accurately (e.g., by agarose gel electrophoresis and use of proper mass markers), the presence of external DNA affects data quality. One distinct possibility is the potential presence of additional priming sites that can result in double priming. In addition, the presence of huge molecules of chromosomal DNA may create unfavorable reaction conditions and interfere with the electrokinetic injection in capillary instruments. Figure 2.6 shows the effect of DNA and RNA

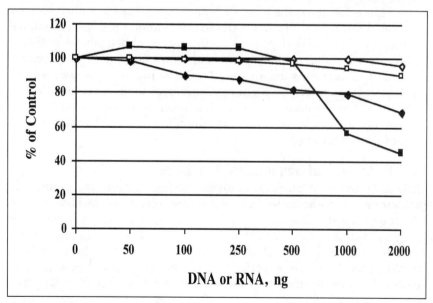

Figure 2.6. The effect of foreign DNA and RNA. Plasmid DNA (pGem3zf) was heat-denatured for 7.5 minutes at 98°C in the presence of varied concentrations of DNA (calf thymus; purchased from Invitrogen, Carlsbad, CA) or RNA (total RNA from human tissues; purchased from Invitrogen). The molecules were absent (■) DNA and (□) RNA, or present (◆) DNA and (◇) RNA during heat-denaturation.

present during heat-denaturation or added after this step. At DNA concentrations below 0.25 µg/reaction (25 µg/ml), there is little effect on read lengths and quality regardless of whether external DNA was present or absent during the denaturation step. However, at concentrations above 1 µg/reaction (100 µg/ml) DNA added after the denaturation step has a much stronger inhibitory effect. At 2 µg/reaction (200 µg/ml) inhibition is over 50% versus about 30% when the foreign DNA was present during the denaturation step. On the other hand, RNA (at the amounts up to 2 µg/reaction = 200 µg/ml) does not affect sequencing outcome regardless of whether it was present or absent during the heat denaturation step.

Effects of Primer Concentration

In a typical DNA sequencing protocol, 20 to 25 cycles are used to produce adequate signal strength for optimal read length. Therefore, the molar ratio of primer to template should be at least 20 to 25. For example, using a 25-fold ratio of primer over template and setting up the thermocycler for 30 to 35 cycles should not improve data quality and read length. Similarly, using a 50-fold molar excess of primer over template and performing 25 cycles should not have any effect on the quality and read length. If it does, as some workers have observed (2), then either the concentration of template or primer (or both) may have been determined incorrectly. Figure 2.7 shows the effect of primer concentration (either in crude or purified form, using a reverse-phase purification protocol) on optimal read length. For purified primers, using more primer (above a 25-fold molar ratio of primer over template, with 25 sequencing cycles) does not result in increased read length. However, 20% inhibition is seen when a substantial molar excess of a crude primer over template is used, which results from the presence of salts or other impurities carried over from the oligo synthesis. Similar data were obtained for many different combinations of DNA primer (not shown).

Troubleshooting of Sequencing Reactions

Table 2.1 provides a practical guide to solving some of the sequencing problems one can encounter in a typical core sequencing facility. Usually, the sequencing staff has no control over which DNA isolation method is being used and the quality/quantity of submitted DNA preparations; it is not unusual that the DNA concentration is off by a factor of 5- to 50-fold. Therefore, the experience and the first-hand knowledge of "customers" are paramount to the success and efficiency of a particular sequencing unit.

This troubleshooting guide covers mainly the issues arising from the "biochemical" part of the DNA sequencing process. The manuals that typically accompany any DNA sequencing instrument cover in sufficient detail the technical side of equipment related to sequencing problems.

Chapter 2. Effects of Various Reagents

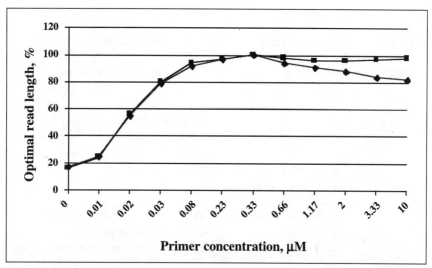

Figure 2.7. The effect of primer concentration. A 0.25 μg/reaction of pGem3zf plasmid DNA was heat-denatured for 7.5 minutes at 98°C in the presence of varied concentrations of RP-1 column purified (■) or crude (♦) M13 forward primer. At 0.33 μM of primer, the molar ratio of primer to DNA is approximately 25 under this experimental condition.

Summary

DNA quality is the single most important factor in obtaining high quality sequence data. By paying careful attention to DNA purification protocol(s), quantity of the template, and the potential presence of common contaminants, one can greatly increase quality and read length. In troubleshooting poor results, it may be very helpful to know the "history" of the template: how it was prepared, quantified, and stored. Often the source of poor quality data arises from confusion over the precise meaning of names for common reagents. For example, for some "TE" means 10 mM Tris-HCl/1 mM EDTA, pH 8.0, but for others it may mean 10 mM Tris-HCl, 0.1 mM EDTA pH 7 to 8. This confusion may have a negative impact on sequencing data, especially when a very diluted DNA sample is submitted for processing. The data in this chapter can be used as a guide to the most optimal conditions for setting up and running DNA sequencing reactions in any core industrial or academic center. It is practically impossible to impose a set of standard protocols for preparation, quantification, and storage of templates, and, therefore, the knowledge of factors influencing sequencing outcome may help to pinpoint the problem and suggest solution.

Table 2.1. Troubleshooting.

Problem	Likely Cause	Possible Solution(s)
No data	No or not enough DNA	• Verify the DNA concentration by gel or UV and adjust accordingly. Use the recommended amount of DNA: for plasmid DNAs use 0.1–1.0 µg and for PCR products use 0.05–0.2 µg/reaction.
	More than 4 µg of DNA/reaction used	• See above.
	Information about primer or vector is incorrect	• Verify vector information and correct, if necessary. • Verify that primer number (name or whatever tracking system for primers is being used) matches that on printout sheet. • Verify that correct version of vector was used for cloning (e.g., priming site was removed). Select different set of primers.
	Suspect or bad primer	• Consult the specs sheet accompanying the delivered primer. Contact primer provider for additional data. Alternatively, run an aliquot of a suspected primer on 15–20% acrylamide gel. Look for N-1 bands and lower than specified concentration. If needed, ask for another synthesis of the same primer.
	Thermocycler issues	• The wrong cycling program was selected. Repeat the sequencing with the correct settings. • The correct program was selected but not activated. Activate the cycling program.
	Problems with the purification of sequencing reactions and	• Make sure that the correct solutions, resin, and volumes were used to prepare the sieving columns (for G-50–like method).

Table 2.1. *Continued*

Problem	Likely Cause	Possible Solution(s)
	all labeled reactions were lost.	Keep and record the lot numbers of all components used in this step. • Make sure that the correct concentrations of salts and alcohol were used to precipitate sequencing reactions. Keep and record the lot numbers of all components used in this step. • If using a commercial kit, consult with its provider for possible suspect lot numbers. Keep and record the lot numbers of all components used in this step.
Poor quality data	Not enough DNA	• Verify the DNA concentration by gel or UV and adjust accordingly. Use the recommended amount of DNA: for plasmid DNAs use 0.1–1.0 µg and for PCR products use 0.05–0.2 µg/reaction.
	Poor quality DNA (too much salt, EDTA, foreign DNA, or degradation, etc.)	• Confirm the specifics of template preparation. Suggest re-precipitation or ask for the template to be prepared using other proven methods.
	Suspect or bad primer	• Consult the specs sheet accompanying the delivered primer. Contact primer provider for additional data. Alternatively, run an aliquot of a suspected primer on a 15–20% acrylamide gel. Look for N-1 bands or lower than specified concentration. If needed, ask for another synthesis of the same primer.
	Double priming	• Consult the specs sheet accompanying the delivered primer. Contact primer provider for additional data. A strong N-1 band or doublets will cause severe double priming.

Table 2.1. Continued

Problem	Likely Cause	Possible Solution(s)
	Problems with the purification of sequencing reactions and part of labeled reactions were lost.	• Check if the primer has more than one annealing site. Design a new primer, if needed. • The primers were not removed from the PCR fragments. Repurify. • Make sure that the correct solutions, resin, and volumes were used to prepare the sieving columns (for G-50–like method). Keep and record lot numbers of all components used in this step. • Make sure that the correct concentrations of salts and alcohol were used to precipitate the sequencing reactions. Keep and record the lot numbers of all components used in this step. • If using a commercial kit, consult with its provider for possible suspect lot numbers. Keep and record lot numbers of all components used in this step.
	Thermocycler issues	• Non-optimal cycling program was selected. Redo and rerun sequencing reactions. • Using a cycling block with possible problems (e.g., heat uniformity failure). Redo and re-run the reactions.
Short run	Buffer leaked after few hours of the run (this is especially relevant for instruments with slab gel technology: ABI373/377).	• If the read length is satisfactory use the data. On the next run make sure that the clamps are tightened evenly. • Check the rubber seal on the upper buffer tank for possible damage. Replace, if necessary. • Inspect the syringes and/or tubing on the ABI 3100/3700/3730 and adjust, if necessary.

Table 2.1. *Continued*

Problem	Likely Cause	Possible Solution(s)
	ABI instrument did not recover after a power failure	• Rerun samples.
	Run time and Sheet flow run time do not match (ABI3700)	• Make sure that the Run time matches the Sheet flow run time. Consult ABI manual. For the default run module these parameters should be 10,000 sec and 12,000 counts, respectively.
Data abruptly end after about 150–250 bases (peaks off scale in the readable region).	Too much DNA (amount of DNA/reaction between 2–3 µg)	• Verify the DNA concentration by gel or UV and adjust accordingly. Use the recommended amount of DNA: for plasmid DNAs use 0.1–1.0 µg and for PCR products use 0.05–0.2 µg/reaction.
Data abruptly end after long GC stretch.	GC content above 70%	• Use any published protocols that deal with difficult templates. 1 M betaine and various additives (e.g., SEQUENCERx Enhancer System from Invitrogen) included in the reaction is a good start. See also Chapter 3 in this book.
Data with broad peaks throughout the entire sequence	Too much DNA (amount of DNA close to 4 µg/reaction)	• Verify the DNA concentration by gel or UV and adjust accordingly. Use the recommended amount of DNA: for plasmid DNAs use 0.1–1.0 µg and for PCR products use 0.05–0.2 µg/reaction.
G- or other spikes in the sequencing data	Bubbles in the polymer	• Check the delivery tubing on capillary instruments. If bubbles are visible, flush them out and rerun the affected samples.

Table 2.1. *Continued*

Problem	Likely Cause	Possible Solution(s)
	Expired polymer	• Replace the polymer with a fresh un-expired polymer lot.
	Possible crystals or contaminant in the polymer	• Adjust the temperature of the polymer to room temperature (do not heat to thaw rapidly). Gently swirl to dissolve any possible solids.
Unidentified		• Rerun the affected samples on a different instrument. If the G-spikes are still present, redo all reactions.
Beginning of the sequence with huge T- or other blobs	Problems with the purification of sequencing reactions	• Make sure that the correct solutions, resin and volumes were used to prepare the sieving columns (for G-50–like method). Keep and record the lot numbers of all components used in this step. • Make sure that the correct concentrations of salts and alcohol were used to precipitate the sequencing reactions. Keep and record the lot numbers of all components used in this step. • If using a commercial kit, consult with its provider for possible suspect lot numbers. Keep and record lot numbers of all components used in this step.
	Too dilute Taq mix was used.	• Check if the Taq mix was properly diluted. Visually compare to other dilutions. Make a new dilution or use a different one and repeat sequencing.
Sequence data does not at all match with provided reference sequence.	Wrong reference sequence was used. Wrong clone was sent.	• Verify that the correct reference sequence was used and re-analyze the sequencing data. • Verify that the correct clone was used and re-analyze the sequencing data.

Chapter 2. Effects of Various Reagents

Table 2.1. *Continued*

Problem	Likely Cause	Possible Solution(s)
Your sequence has some differences with provided reference sequence.	Reference sequence was based on genomic or cDNA sequence.	• Make sure that the quality of your data is high and you have double-stranded coverage.
	Reference sequence was "pieced together" from various electronic sequences.	• Make sure that the quality of your data is high and you have double-stranded coverage.
Forward and reverse strands have one or more base differences.	Sequencing on different strands was done on two separate isolates of clones.	• Redo the entire sequencing on the same isolate, if possible.
	If series of similar clones were submitted, sequencer accidentally mixed two clones.	• Redo the entire sequencing on the same clone(s).
Other run problems	Various	• Please refer to "Run problems" in the "Troubleshooting" sections of the relevant DNA sequencing instrument manuals.

References

1. ABI PRISM® BigDye™ Terminator v3.0 Ready Reaction Cycle Sequencing Kit. *Protocol*. 2001. Applied Biosystems. Foster City, CA.

2. Azadan, R.J, Fogleman, J.C., and Danielson, P.B. 2002. Capillary electrophoresis sequencing: maximum read length at minimal cost. *BioTechniques* 32: 24–28.
3. Fuller, C.W. 1992. Modified T7 DNA polymerase for DNA sequencing. *Methods Enzymol* 20: 329–354.
4. Tabor, S., and Richardson, C.C. 1989. Effect of manganese ions on the incorporation of dideoxynucleotides by bacteriophage DNA polymerase and *E. coli* DNA polymerase I. *Proc Natl Acad Sci USA* 86: 4076–4080.
5. Taylor, G.O., and Dunn, I.S. 1994. Automated cycle sequencing of PCR templates: relationships between fragment size, concentration and strand renaturation rates on sequencing efficiency. *DNA Seq* 5: 9–15.
6. T7 Sequenase V2.0. 1997. T7 Sequenase version 2.0 DNA sequencing kit. Amersham Life Science. Little Chalfont, Buckinghamshire, UK.

3 Sequencing of Difficult DNA Templates

Jan Kieleczawa
Wyeth Research, Cambridge, MA

Introduction

The release of a comprehensive draft of the human genome sequence in June 2000 (14, 22) provided the scientific community with one of the greatest tools in history to advance basic and applied knowledge about the fundamental aspects of human biology and the molecular basis of genetic diseases. In addition, comparative sequencing of other model organisms (yeast, nematode, fruit fly, rat, mouse, dog, etc.) already has shed immeasurably valuable information on a number of issues, and will continue to do so as the sequences for more organisms become available.

It is relatively easy to sequence even a large and complex genome in draft form. However, it took almost two years and tremendous effort on the part of several large sequencing centers to officially declare the human genome as completed in April of 2003. Some of the reasons for that delay were the DNA fragments in question were unclonable and highly repetitive (and hence, difficult to assemble), or belonged to the category of so-called "difficult templates." It is safe to assume that similar issues will be encountered in finishing other complex genomes.

This chapter describes a number of protocols that have potential for sequencing many kinds of difficult templates. For the purpose of this review we consider a DNA entity as a "difficult template" if it cannot be sequenced using the standard DNA sequencing protocol (1). We also assume that templates are of high quality and free of any potential inhibitory contaminants (see also Chapter 2). The difficult templates can be classified into (at least) the following categories:

1. GC-rich (over 65% GC);
2. Containing di- or more nucleotide repeats (for example CTT, CCT, TTTCCC,CCA, GCC, GA, GT) and any combination of these repeats;

3. Alu repeats;
4. Other repeats (direct and inverted);
5. Strong hairpins;
6. PolyA/polyT or other homopolymer regions

The most common technique used to help to sequence through the GC-rich regions is to add DMSO to final concentration of 5% (2–4, 7, 13, 17, 19). Recently, some progress in the sequencing of difficult templates was made by adding to the sequencing mix a number of additives (betaine, one of seven sequence enhancer Rx reagents from Invitrogen), using modified Taq chemistries like the ABI cycle sequencing dGTP kit (8) or the newest BigDye terminators 3.1 and 1.1 (6). New developments from Amersham Biosciences (16) also must be acknowledged. Hopefully, some of this research will be published in the form of scientific papers. Some success in sequencing templates belonging to categories 1/2/4/5 can also be realized by mixing (4:1 vol/vol) BigDye terminator 3.0/3.1 with dGTP (V3.0) dye-terminator (George Grills, personal communication) or adding Nonident P-40 detergent (Bruce Roe, personal communication). In some cases, using a modified cycling protocol proved to be effective (8). Using a hot sample loading technique, Gerstner et al. (11) were able to get through a GC-rich region manifesting as a compression under standard run conditions. Similarly, substitution of dATP with 7-deaza-dATP appeared to help sequence through a compression caused by a 5'-YGN$_{1-2}$AR motif where Y and R are base-pairing pyrimidine and purine bases, respectively (24). Although transposon technology (5, 12) requires additional labor-intensive steps, it appears to be effective in sequencing through some difficult templates (18; JK, unpublished results). Limited success with long polyA/T tracts can be accomplished using polyT/polyA primers with a degenerate base at the 3' end (20, JK, unpublished results). A more elegant approach was employed by Langan et al. (15), who used primers that were adjacent to or contained (at the 5' end) part of the polyA/polyT tracts. This resulted in greater than 99% correct base calling accuracy.

In recent years RNAi technology has become a powerful tool to explore gene function (21 and references therein). One approach in mammalian cells is to use siRNA vector plasmids to express sense and antisense regions separated by a loop. Such vectors form hairpin structures in the DNA that may interfere with sequencing under standard conditions. Ducat et al. (9) devised a technique that includes using digestion with a specific restriction enzyme in the hairpin to remove its secondary structure. This approach requires that a restriction site be present at a specific location for effective relief of tension in the secondary structures. This could be incorporated in the design of a particular hairpin structure.

Chapter 3. Sequencing of Difficult DNA Templates

In this chapter, we evaluate several protocols that can be used to sequence various difficult templates: we primarily concentrate on GC-rich templates and those that contain various di- or more nucleotide repeats. Each template was sequenced using almost all protocols described below. All of these protocols take advantage of controlled heat-denaturation of the plasmids (see Chapter 1) before addition of Taq sequencing mix. This difference alone was sufficient to sequence through a fragment of DNA in vectors containing various hairpin structures or inverted repeats without the need for restriction enzyme digestion.

Materials and Methods

Materials

All fluorescent dye-terminators were purchased from Applied Biosystems (Foster City, CA). Betaine was bought from Sigma-Aldrich (St. Louis, MO). A set of seven sequence enhancer Rx reagents was from Invitrogen (Carlsbad, CA). Thermofidelise was from Fidelity Systems (Gaithersburg, MD). GC-melt reagent was purchased from BD Biosciences (San Jose, CA). All other reagents were of the highest available purity.

All additives, except for Thermofidelise, were added before heat denaturation. There were no differences when all additives were added after heat denaturation (JK, data not shown). All sequencing reactions were run on an ABI3700 or an ABI3100 equipped with 80-cm capillaries (Applied Biosystems). Each data point is the average of at least three reactions. The data are presented as the longest reads with phred quality ($Q \geq 20$). In dealing with any difficult region, the sequencer's intervention is often critical (visual inspection of the data, editing), and it is more important to get through the region than to get data of the highest quality.

The plasmid DNAs used in these tests were of the highest possible quality and were prepared using common and proven DNA purification protocols. We have tested various modifications of the standard ABI protocol (1) to sequence through difficult regions.

The following list contains all of the protocols tested in this work. They are referred to by number in Table 3.1.

1. Protocol #1 (the standard protocol for this review). In a 0.2 ml tube mix x µl of DNA (final amount = 250 µg) with 1 µl of 5 µM primer and y µl of TEsl (10 mM Tris/0.01 mM EDTA, pH 8.0) to a final volume of 7 µl. Add 3 µl of 1.5 diluted BigDye™ BD V3.0 terminator mix (0.8 ml of undiluted Taq FS mix + 0.4 ml of ABI's 5× Dye-terminator buffer = 4× final Taq FS dilution) and start standard cycle sequencing (5 sec at 50°C/4 min/60°C/10 sec at 96°C) × 25.

Table 3.1. Characteristics of difficult templates.

DNA Number	C:G	Method Number → Characteristics ↓	1	2	3	4	5	6	7	8	9	10	11	12	13
1	NA	CCT/TTTCCC-repeats	222	277	193	898	914 (B/D)	678	341	477	920	933	NT	NT	NT
2	1.00	TCC/GCC-repeats	0	830	853	848	929 (C)	893	689	855	888	852	NT	NT	NT
3	1.82	CTT-repeats	401	431	411	466	540 (BDE)	442	457	455	460	465	430	426	NT
4	NA	Alu-repeats	0	118	107	244	262 (A)	258	111	463	820	801	NT	NT	NT
5	0.75	GC-rich/90%	0	267	270	288	304 (A)	302	283	318	505	565	270	420	780
6	0.43	GC-rich/88%	0	504	928	933	967 (C)	988	500	749	895	976	910	675	NT
7	1.18	GC-rich/75%	0	660	960	940	974 (F)	869	520	747	932	914	753	713	NT
8	1.54	GC-rich/67%	0	989	222	1002	1087 (C)	1025	145	177	361	201	467	945	NT
9	1.09	GC-rich/63%	208	572	121	557	673 (C)	535	211	562	421	611	740	549	NT
10	NA	Plasmid with RNAi hairpin-R primer	0	590	601	885	710 (A/D)	786	548	758	811	861	656	811	NT
11	NA	Plasmid with RNAi hairpin-F1 primer	566	820	750	723	806 (C/D)	806	635	930	963	992	775	955	NT
12	NA	Plasmid with RNAi hairpin-F2 primer	548	685	514	638	766 (C/D)	796	586	826	855	817	657	589	NT

Column 3 lists major characteristics of each template. When relevant, the C:G ratio is indicated in column 2. In each template, this ratio is for the "difficult" region only. When one of the simpler treatments resulted in a good-quality sequence then more elaborate protocols were not tried (NT). The data are expressed in terms of phred Q ≥ 20 values. Templates 10–12 had similar hairpin structures, and one reverse (R) and two different forward primers (F) were used for sequencing.

Chapter 3. Sequencing of Difficult DNA Templates

2. Protocol #2. In a 0.2 ml tube mix x μl of DNA (final amount = 250 μg) with 1 μl of 5 μM primer and y μl of TEsl (10 mM Tris/ 0.01 mM EDTA, pH 8.0) to a final volume of 7 μl. Heat-denature at 98°C for 5 minutes (see Chapter 1 for details). Place the tube(s) on ice, add Taq terminator mix, and cycle as in protocol #1.
3. Protocol #3. Like protocol #2 but DMSO added to a final concentration of 5%.
4. Protocol #4. Like protocol #2 but betaine (1 M final concentration) was added.
5. Protocol #5. Like protocol #2 but one of seven Rx reagents was added.
6. Protocol #6. Like protocol #2 but GC-melt (1x final) was added.
7. Protocol #7. Like protocol #2 but 0.25 μl of Thermofidelise (V1.0) was added.
8. Protocol #8. Like protocol #2 but dGTP (V3.0) dye-terminator kit was substituted for regular, V3.0, dye-terminator.
9. Protocol #9. Like protocol #8 but betaine (1 M final concentration) was added.
10. Protocol #10. Like protocol #8 but reagent A was added to a final concentration of 1×.
11. Protocol #11. Like protocol #2 but dye-terminator V3.1 was used instead of V3.0.
12. Protocol #12. Like protocol #2 but a 4:1 (vol/vol) mixture of BD V3.1: dGTP (V3.0) was added.
13. Protocol #13. This is a two step protocol that can be used to sequence GC-rich templates. In the first step the G-rich DNA fragment is PCRed out using primers flanking the DNA region in question with a complete substitution of dGTP with 7-deaza-dGTP (10). The PCR fragment is run on a 1% agarose (23) gel to confirm the presence of the expected product and then purified using one of the standard protocols (for example, GFX cartridges from Amersham Biosciences, Piscataway, NJ). Please note that the amount of DNA produced using deaza-dGTP is about 20% of similar fragments prepared with dGTP (10). In step 2, the purified product is then sequenced using protocol #1 (no need for heat denaturation; however, 30–60 second heat-denaturation can be used. JK, unpublished results).

Results

Table 3.1 shows the data for the longest reads expressed with $Q \geq 20$ phred values for various difficult templates. In method #5, all seven reagents from Invitrogen's set were tested and only the best data are presented here, with indicated reagent's name. If several reagents gave similar results it is noted in this column. For DNA #5, the only way to sequence through this 90% GC-region was to apply method #13. For this particular DNA, as well as for DNA #3, a number of different, commercially available, preparation methods were used; however, no improvements in read length were seen (JK, data not shown). The data presented in Table 3.1 refers to a particular set of DNA preparations and sequencing conditions. It is entirely possible that under a different set of conditions (type of DNA preparation, chemistry, instrumentation, etc.) the numeric values for the data would be different from those presented here.

Discussion

In all cases the controlled heat-denaturation of plasmids, without any other treatment, produced sequence data of much better quality and increased read length. In seven out of twelve cases (58%), this was the only way that any data could be produced. When using the standard sequencing protocol (protocol #1), DMSO is known to help to sequence through some GC-rich regions (see introduction for references). However, if a heat-denaturation step that produces an effective strand separation is part of the sequencing protocol, DMSO has no, or very little, effect on sequence quality. In fact, in some cases the addition of DMSO was inhibitory and we do not recommend its addition when a heat-denaturation step is used. Adding betaine or reagent "A" increased the read length in most cases and these reagents seem to be equally effective. Betaine, however, is a more economical alternative. Reagents "B/C/D" seem to be most effective with C-rich templates. The addition of Thermofidelise™ was not beneficial or inhibitory under our experimental conditions. The substitution of dye-terminator dGTP™ 3.0 for BigDye™ V3.0 with BD was successful in 58% of the tests (especially with G-rich templates). Substituting BigDye™ V3.1 for BigDye™ V3.0 had only limited success (20%) and half of the reactions (50%) produced the same read length when using the heat-denaturation step with BigDye™ V3.0. Adding the combination of BD 3.1:dGTP V3.0 (vol/vol 4:1) was effective 45% of the time. However, 55% of the time the data were the same or worse compared to the protocol using the simple heat-denaturation step. Other treatments, for example, adding GC-melt reagent, gave slightly better read lengths in some cases, but did not offer any substantial advantage over using betaine or one of the seven reagents from Invitrogen's set.

The presented data are based on a limited number of templates belonging to different categories of difficulty. As a greater number of various difficult templates is accumulated, we will be able to expand the scope of this study with the goal of developing general protocols for dealing with different types of difficult templates that will have the best probability of success. Sometimes, redoing the DNA preparation or

Chapter 3. Sequencing of Difficult DNA Templates

switching to an entirely different protocol may be helpful (JK, unpublished observations). We realize, however, that it may be impossible to fully predict which protocol to use even with *a priori* knowledge of the type of a difficult template. A trial-and-error approach may still be the only effective way to solve a particular sequencing issue. We hope, however, that the data presented in this Chapter will serve as a useful guide for a more systematic approach to deal with unusual sequencing situations.

References

1. ABI PRISM® BigDye™ Terminator v3.0 ready Reaction Cycle Sequencing Kit. 2001. Protocol. Part number 4390037 Rev. A. Applied Biosystems. Foster City, CA.
2. Adams, P.S., Dolejsi, M.K., Hardin, S., et al. 1996. DNA sequencing of a moderately difficult template: Evaluation of the results from a *Thermus thermophilus* unknown test sample. *BioTechniques* 21: 678.
3. Adams, P.S., Dolejsi, M.K., Hardin, S., et al. 1997. Effects of DMSO, thermocycling and editing on a template with a 72% GC rich area: results from the 2nd Annual ABRF Sequencing Survey demonstrate that editing is the major factor for improving sequencing accuracy. Ninth International Genome Sequencing and Analysis Conference. *Microb Comp Genomics* 2: 198 (abstract).
4. Adams, P.S., Dolejsi, M.K., Grills G., et al. 1999. An analysis of techniques used to improve the accuracy of automated DNA sequencing of a GC-rich template: results from the 2nd ABRF DNA Sequence Research Group Study. Available at: http://www.abrf.org, search: 2nd ABRF DNA Sequence Research Group Study. Accessed January 2004.
5. Bhasin, A., Goryshin, I.Y., and Reznikoff, W.S. 1999. Hairpin formation in Tn5 transposition. *J Biol Chem* 274: 37021–37029.
6. Bond, D., Swei, A., Lee, K., Nutter, R., et al. 2002. New advances in DNA sequencing chemistry. Genome Sequencing and Analysis Conference, Boston. B-01, 57 (abstract).
7. Burgett, S.G., Rosteck Jr., P.R. 1994. Use of dimethyl sulfoxide to improve fluorescent, Taq cycle sequencing. In Adams, M.D., Fields, C., and Venter, J.C., eds. *Automated DNA Sequencing and Analysis*. Academic Press. NY, 211–215.
8. Chapter 7. *Automated DNA Sequencing. Chemistry Guide* (document number 4305080B). 2000. Applied Biosystems. Foster City, CA.
9. Ducat, D.C., Herrera, F.J., and Triezenberg S.J. 2003. Overcoming obstacles in DNA sequencing of expression plasmids for short interfering RNAs. *BioTechniques* 34: 1140–1144.
10. Fernandez-Rachubinski, F., Murray, W.W., Blajchman, M.A., and Racubinski, R.A. 1990. Incorporation of 7-deaza dGTP during the amplification step in the polymerase chain reaction procedure improves subsequent DNA sequencing. *DNA Seq* 1: 137–140.

11. Gerstner, A., Sasvari-Szekely, M., Kalasz, H., and Guttman, A. 2000. Sequencing difficult DNA templates using membrane-mediated loading with hot sample application. *BioTechniques* 28: 628–630.
12. Goryshin, I.Y., and Reznikoff, W.S. 1998. Tn5 in vitro transposition. *J Biol Chem* 273: 7367–7374.
13. Hawes, J.W., Escobar, H., Hunter, T., et al. (2003) Sequencing Through Difficult Repetitive Sequence. Study 2003. Results from the ABRF DNA Sequence Research Group Study. Available at http://www.abrf.org, search: DSRG 2003 Study. Accessed January 2004.
14. International Human Genome Sequencing Consortium. 2001. Initial sequencing and analysis of the human genome. *Nature* 409: 860–921.
15. Langan, J.E., Rowbottom, L., Liloglou, T., et al. 2002. Sequencing of difficult templates containing poly(A/T) tracts: closure of sequencing gaps. *BioTechniques* 33: 276–280.
16. McArdle, B., Kerelska, T., Marks, A., et al. 2002. New sequencing chemistry and analysis software improve the Megabace DNA analysis systems. Genome Sequencing and Analysis Conference. Boston. B-14, 60 (abstract).
17. Naeve, C.W., Buck, G.A., Niece, R.L., et al. 1995. Accuracy of automated DNA sequencing: a multi-laboratory comparison of sequencing results. *BioTechniques* 19: 448–453.
18. Rogosin, A., Rubenfield, M., Smith, D., et al. 2003. Transposons: an alternative to primer walking and a strategy for closing difficult gaps. Advances in Genome Biology & Technology in cooperation with Automation in Mapping & DNA Sequencing. Marco Island, FL (abstract #100).
19. Seto, D. 1990. An improved method for sequencing double stranded plasmid DNA from minipreps using DMSO and modified template preparation. *Nucleic Acid Res* 18: 5905–5906.
20. Thomas, M.G., Hesse, S.A., McKie, A.T., and Farzaneh, F. 1993. Sequencing of cDNA using anchored oligo dT primers. *Nucleic Acids Res* 21: 3915–3916.
21. Tuschl, T. 2002. Expanding small RNA interference. *Nat Biotechnol* 20: 446–448.
22. Venter, J.C., Adams, M.D., Myers, E.W., et al. 2001. The sequence of human genome. *Science* 291: 1304–1351.
23. Vogelstein, B., and Gillespie, D. 1979. Preparative and analytical purification of DNA from agarose. *Proc Natl Acad Sci USA* 76: 615–619.
24. Yamakawa, H., and Ohara, O. 1997. A DNA cycle sequencing reaction that minimizes compressions on automated fluorescent sequencers. *Nucleic Acid Res* 25: 1311–1312.

4 New DNA Sequencing Enzymes

Sandra L. Spurgeon and John W. Brandis
Applied Biosystems, Foster City, CA

Introduction

In the 27 years since the publication of Sanger's work (56), the chain-termination method for DNA sequencing utilizing dideoxynucleotides has become one of the most widely used methods in molecular biology. In that time the technology has changed from a low-throughput, manual, radioactive method to one that has been adapted to the high throughput demands required to sequence the entire human genome (30, 64). This dramatic advance has required substantial improvements in all aspects of the sequencing process (43, 44), but especially in the development of automated sequencing instruments and fluorescent sequencing methods.

The first commercially available automated DNA sequencer was the ABI Model 370, introduced in 1987 (14). This was a slab gel-based instrument and its throughput for a five-day work week was about 30,000 bases (23). Today, with a high throughput instrument like the ABI Prism® 3730xl Genetic Analyzer, a 96-capillary machine, the throughput approaches 2 million quality (Q20) bases per day. Because this instrument also can run unattended over weekends, approximately 9,500,000 bases/instrument can be generated in a five-day work week. This is an increase of more than 300-fold in the number of bases per instrument per week.

In addition to the dramatic increase in throughput, data quality has improved significantly. A very important part of this has been the development of better sequencing chemistries. The introduction of new fluorescent dyes, optimized sequencing protocols, and modified sequencing enzymes all have contributed to the improvement of the sequencing chemistries. With current sequencing chemistries, the read length is frequently more a function of the resolution of the detection system rather than any limitation on the part of the sequencing chemistry. However, with more emphasis on the use of DNA sequencing for such demanding applications as forensics and diagnostics, there continues to be a need for further improvements in the quality and reliability of data. Because DNA polymerases play a central role in dideoxysequencing, the development of new and better enzymes is expected to be very important. The purpose of this chapter is to discuss recent work at Applied Biosystems in the development of new and modified polymerases for automated sequencing using fluorescent dyes on the ABI Prism instruments. To provide a basis for that discussion, a brief review of DNA polymerases and the development of current sequencing enzymes are included.

DNA polymerases are enzymes that catalyze the synthesis of DNA and are required for DNA replication and repair (17, 36). At least six classes of polymerases are currently recognized (51). Families A, B, and C are based on the amino acid sequence homology to *E. coli* Pol I, Pol II, and Pol III (6, 31). Other families are X, RT, and UmuC/DinB. The last group has also been referred to as the Y-family (24). The only enzymes that have been used extensively for DNA sequencing applications are members of family A, in particular, modified forms of T7 DNA Polymerase and *Taq* Pol I. The polymerase most widely used for automated DNA sequencing and that was largely responsible for sequencing the human genome is *Taq* Pol I.

The first enzymes to be used for automated sequencing were the same enzymes that were used for radioactive sequencing at that time: Klenow fragment of *E. coli* Pol I and modified T7 DNA polymerase (14, 23, 32). In 1988 a new enzyme, *Taq* Pol I, became available for use in both polymerase chain reaction (PCR) (55) and sequencing (29). Although *Taq* Pol I was initially used in manual radioactive methods, protocols were soon developed for automated sequencers as well (16, 25). The use of this enzyme resulted in significantly higher peak intensities and better peak height uniformity for dye-labeled primer chemistries. The sequencing process was improved further with the development of cycle sequencing protocols that took advantage of the thermostability of *Taq* Pol I (11, 47). The use of cycle sequencing protocols are advantageous in that they reduce the number of steps required to carry out sequencing reactions, double-stranded templates can be sequenced as easily as single-stranded ones, and the amount of template in the sequencing reaction can be reduced (15).

Another key innovation in fluorescent sequencing was the development of dye-labeled terminators suitable for use on the Applied Biosystems platforms. In 1987 Prober et al. (53) assembled a set of fluorescently labeled dideoxynucleotides for use on the DuPont Genesis 2000. However, this dye set was not suitable for use on the ABI 370 DNA Sequencer, which required greater spectral resolution. In 1990, a set of dideoxynucleotides labeled with rhodamine dyes became available that could be used with *Taq* Pol I to carry out four-color automated DNA sequencing (4). The use of labeled dideoxynucleotides was advantageous in that only fragments terminated with a dideoxynucleotide were labeled

Chapter 4. New DNA Sequencing Enzymes

and all four terminator reactions could be done in a single tube. However, the peak height profiles generated by these dye-labeled terminators were less uniform than the traces for dye-labeled primer chemistries (48). For this reason, many of the high throughput genome sequencing laboratories continued to use dye primer chemistry despite the advantages of using dye labeled terminators.

Development of AmpliTaq DNA Polymerase FS and Other Improvements

The growing importance of automated sequencing in the early 1990s and the need to improve this technology resulted in efforts to find better sequencing enzymes at Applied Biosystems. Native, full-length *Taq* Pol I has an active 5'-3' structure specific nuclease that resides in a separate domain from the polymerization site. One of the first areas we investigated was the impact of the 5'-3' nuclease activity on cycle sequencing. This activity can be removed either by truncation, such as in Stoffel fragment in which the nuclease domain has been removed entirely (37), or by site-directed mutagenesis. Data from dye primer reactions generated with enzymes that were missing the 5'-3' nuclease activity showed reduced background noise and better quality (60). Although the truncated Stoffel fragment was shown to be more thermostable (37), the preferred enzyme was a point-mutant of full-length enzyme, namely *Taq* G46D, which retained the processivity and rate of extension of the full-length enzyme. Additional studies showed that the occasional loss of peaks at certain positions in data generated with *Taq* dye primer chemistry was due to the reversal of polymerization (pyrophosphorolysis) as inorganic pyrophosphate begins to accumulate during the course of the reaction. Tabor and Richardson (62) showed that including a pyrophosphatase along with T7 DNA polymerase could restore a similar loss of peaks. Hence, *Taq* G46D was formulated with a thermostable inorganic pyrophosphatase and this enzyme was commercialized as AmpliTaq DNA Polymerase CS initially for dye primer chemistry.

One of the major problems still remaining with AmpliTaq (a registered trademark of Roche Molecular Systems) DNA Polymerase CS was its strong bias against the incorporation of dideoxynucleotides (9). This behavior differed significantly from the behavior of T7 DNA Polymerase, which showed very little bias against the incorporation of dideoxynucleotides and also gave much more even peak heights in dye primer sequencing (61).

Attention, therefore, was focused on trying to eliminate the bias against dideoxynucleotides shown by *Taq* Pol I to improve peak height uniformity. Tabor and Richardson (63) found that replacement of phenylalanine by tyrosine at a single amino acid position—667—in *Taq* Pol I virtually eliminated the bias against dideoxynucleotides. The same mutant was found by Kalman (33) who used a nucleotide selectivity screening procedure (28). When F667Y was combined with G46D, the resulting enzyme showed very little 5'-3' nuclease activity (1/3000th of wildtype; unpublished observations) and no discrimination against dideoxynucleotides (9). This enzyme also showed much more even peak heights with dye primer chemistry. The double mutant enzyme was commercialized in 1995 as AmpliTaq DNA Polymerase FS.

Reducing the bias against ddNTPs had other benefits as well. Because of the dramatic reduction in the concentration of terminators required for the sequencing reactions, ethanol precipitation could be used to remove unincorporated terminators. This offered a fast and low cost alternative to the use of spin columns, which was particularly important to high throughput sequencing laboratories.

The development of AmpliTaq DNA Polymerase FS in 1995 was a major advancement for both dye primer and especially dye terminator chemistry. In dye-labeled primer methods, AmpliTaq DNA Polymerase FS produced remarkably uniform peak height patterns. Improvements were observed also in the peak pattern evenness with dye-labeled terminators, although the peak heights were still far from being as uniform as data generated with dye primers (49). This suggests that the linker-arms and dyes must also interact with the polymerase. Clearly, further improvements in dye terminator chemistries would require further changes in the dyes and/or linkers and the enzymes (7).

Efforts to Find Better Dye-Labeled Terminators

When AmpliTaq DNA Polymerase FS was introduced, two dye sets were being used with the enzyme. One set consisting of both fluoresceins and rhodamines (Fam, Joe, Tamra, and Rox) was developed for dye primer sequencing (14). The second dye set consisted only of rhodamines (R110, R6G, Tamra, and Rox) and was developed for dye terminators (4). Both of these sets had similar spectral characteristics and were suitable for use on all existing ABI sequencing platforms. However, as already mentioned, the peak patterns obtained with the rhodamine labeled dye terminators were still uneven despite the F667Y substitution. Furthermore, some background noise that resulted from incomplete correction for the spectral overlap of the dyes (multicomponent noise) was evident in the sequence traces, particularly when the signal was weak. Both of these issues were addressed by the introduction of the d-rhodamine dye terminators. The d-rhodamine dyes are 4,7 dichloro-substituted rhodamines that have narrower emission spectra and better spectral resolution, which reduces multicomponent noise. The AmpliTaq DNA Polymerase FS incorporated the

Chapter 4. New DNA Sequencing Enzymes

dye terminators that were made using these compounds better despite their additional steric bulk (54, 66). The resulting improvement in peak patterns with these dye terminators also was due in part to the development of a new linker arm by which the dye was attached to the nucleotide (34).

The d-rhodamine dyes worked well as single dye terminators or as acceptor dyes for a new class of energy transfer dyes consisting of a Fam as the donor dye and the d-rhodamines as acceptor dyes. The new BigDye (a registered trademark of Biosystems Group, Applera Corp.) Terminators were 2 to 2.5 times brighter with an argon ion laser and had approximately a twofold reduction in multicomponent noise. This resulted in an overall improvement in signal-to-noise of four- to fivefold (38). These new dye terminators were surprisingly advantageous in producing improved peak patterns when used in sequencing reactions with AmpliTaq DNA Polymerase FS. Soon after its introduction in 1997, the BigDye Terminator chemistry became the main sequencing chemistry used on all ABI platforms. Many high throughput laboratories converted to this new chemistry because of its sensitivity, ease of use and high base-calling accuracy.

In 2001 BigDye v3.0 was introduced for use with AmpliTaq DNA Polymerase FS. This chemistry gives more even peak heights in certain sequence contexts and has improved electrophoretic properties for capillary platforms. Although the overall peak pattern evenness is similar to BigDye terminators when dITP is included in the reactions, in the presence of dGTP, the new BigDye v3.0 terminators show remarkably even peak patterns (12).

Efforts to Find Better Enzymes

Today's chemistries are much improved relative to those available for use on the first automated DNA sequencers. In fact, the quality of the data obtained for BigDye Terminators and more recently for BigDye Terminators v3.0 is so high that the limiting factor in automatic base-calling is often due more to fragment resolution rather than to peak height variations for general sequencing applications. However, for certain more rigorous applications such as mutation detection where the correct identification of mixed bases is crucial, the need for more uniform peak height patterns remains. This is particularly important when the ratio of the two bases present is less than 50/50. Also, for *de novo* genomic sequencing, problems with certain contexts still can result in occasional chemistry failures.

Different sequence contexts can affect data quality in different ways. For example, G-rich stretches are especially problematical in causing premature termination in extension reactions because of the presence of dITP in the reaction cocktail. Although dITP is essential for eliminating compressions and is far more effective for that purpose than other agents such as 7-deaza-dGTP, dITP also reduces the stability of duplex DNA (27, 45). For this reason the polymerase can have difficulty extending through G-rich regions. Homopolymeric stretches can result in ambiguous sequence data downstream if they are long enough to result in slippage (2). This is especially true in cDNA sequencing, since various lengths of homopolymer Ts are often found on the 3′ ends of cDNAs. However, such homopolymer stretches also are observed when sequencing genomic DNA. Hence, much attention has been paid to either finding new polymerases or improving the performance of existing enzymes such that they are more tolerant toward modified nucleotides or exhibit reduced amounts of slippage.

Improved polymerases used in automated DNA sequencing must satisfy several minimal requirements. They must be sufficiently thermostable to be used in cycle sequencing protocols that require denaturation temperatures of 95°C or higher. They also must be essentially free of any structure specific nuclease activity (5′ to 3′–nuclease) as mentioned above and free of any 3′ to 5′–exonuclease activity (proofreading). Above all, they must show a low level of discrimination against dideoxynucleotide incorporation. Furthermore, because the most widely used chemistry in automated DNA sequencing is based upon dye-labeled-terminators, any new or improved polymerase must also incorporate these modified dideoxynucleotides well. In addition, the peak height patterns should be at least as uniform as the patterns observed for existing enzyme/chemistry combinations (66).

Sources of New Polymerases

Potential new enzymes for automated DNA sequencing include polymerases from genera other than *Thermus*, enzymes from other *Thermus* species, and, of course, additional modifications to *Taq* Pol I.

Noneubacterial Genera

Hyperthermophilic polymerases from archaeal organisms are widely used for PCR applications because of their very high thermostability and ability to polymerize through sequence anomalies (35). Therefore, enzymes derived from this class of organisms are attractive candidates for DNA sequencing applications (57). However, archael polymerases belong to family B enzymes and share little sequence homology with eubacterial or family A polymerases. Extensive efforts have been made to mutate various archael polymerases to better incorporate dideoxynucleotides,

Chapter 4. New DNA Sequencing Enzymes

but these efforts have met with only limited success (18, 20). While some modifications have been found to improve the incorporation of standard terminators, no single mutation or combination of mutations has the impact of *Taq* F667Y. Recent work with alternative terminators, namely acyclic nucleotides, is promising (21). However, with the standard dideoxynucleotides currently used for fluorescent sequencing, the use of archael polymerases is still not practical.

Other Eubacterial Species: Thermus and Thermotoga

Thermostable DNA Polymerases that are suitable for cycle sequencing have been isolated mainly from eubacterial genera, notably, *Thermus* and *Thermotoga*. At this time, eight species of *Thermus* have been validly described in the literature: *T. aquaticus*, *T. brokianus*, *T. thermophilus*, *T. filiformis*, *T. scotoductus*, *T. oshimai*, *T. antranikianii*, and *T. igniterrae* (13, 46, 65). The previously described *Thermus flavus* AT62 and *Thermus caldophilus* GK24 are now known to be strains of *Thermus thermophilus* (42). Pol I-type enzymes have been purified from several *Thermus* species and a few of these enzymes have been tested in various sequencing applications. Of particular interest in terms of naturally occurring polymerases that may show improved sequence performance properties are the enzymes derived from *T. thermophilus*, a group of halotolerant strains (42) characterized by their ability to grow in 3% salt. The DNA polymerases from two species of *Thermotoga*, *Thermotoga maritima* and *Thermotoga neopolitana*, have been isolated and described (22, 26, 59).

Modified Taq Pol I

While other groups have concentrated more on finding and developing polymerases from other genera, even from other kingdoms, such as the Archaea, our efforts at Applied Biosystems have concentrated mainly on improving the sequencing performance of *Taq* Pol I. As mentioned above, this enzyme was modified by the introduction of two mutations: G46D to reduce the 5' to 3' structure-specific nuclease activity and F667Y to eliminate discrimination against dideoxynucleotides to make AmpliTaq DNA Polymerase FS. We have found that additional mutations, especially in the O-helix region, significantly affected terminator selectivity and sequencing performance both directly through nucleotide binding (7) and indirectly through linker-arm and dye interactions (described later in this chapter).

High-resolution crystal structures show that *Taq* Pol I is a typical family A polymerase and has a so-called "right hand" structure with palm, fingers, and thumb subdomains. In fact, the crystal structures for *Taq* Pol I and the *E. coli* Klenow fragment can be easily superimposed, as shown by Patel and Loeb (50). The thumb region is largely responsible for primer/template binding; the finger region interacts with the in-coming base and undergoes a critical conformational change following correct nucleotide binding that aligns reactive groups in the active site for phosphodiester bond formation (51). Portions of the palm and fingers make up the active site pocket.

One major structural component of the active site and directly involved in nucleotide binding is the O-helix (40, 41). Amino acid sidechains extending from the O-helix interact with the beta- and gamma-phosphates of the in-coming nucleotide, coordinate one of two metal cofactors, and interact with the finger region during the conformational change step (40, 51). Based upon molecular modeling of several different linker-arm/dye structures into various open and closed conformational crystal structures for *Taq* Pol I (unpublished data), we have systematically made amino acid substitutions through site-directed mutagenesis at over 40 different positions in and around the active site (8). The mutation sites were selected for putative enzyme/terminator steric interactions. The mutant enzymes were expressed in and purified from *E. coli* and initially screened for changes in dye-terminator selectivity as described elsewhere (7). Our results revealed two broad classes of interaction sites. Substitutions at amino acid positions located outside of the active site showed only modest selectivity changes that were linker-arm and dye-type specific. Surprisingly, substitutions within the active site, especially along and close to the O-helix, appeared to have far more general effects that were more independent of linker-arm and dye structures. These particular sites are located well away from the linker-arm and dye atoms and are not expected to interact with these structures directly. Nevertheless, these sites strongly influence dye-labeled terminator incorporation. For example, substitutions at R660 biased selectivity in favor of dye-labeled pyrimidines (unpublished data), whereas substitutions at position E681 (located in the "fingers" region) increased the selectivity for dye-labeled purines relative to unlabeled purines. Changes at threonine-664 appeared to "modulate" the effects observed at both positions 660 and 681. We interpret our observations to mean that certain substitutions at these particular sites may cause subtle alterations in the position or interaction of the O-helix within the active site that may in turn affect positioning and interactions of the linker-arms and dyes with the finger and thumb subdomains. In addition to affecting dye-labeled terminator selectivities, substitutions at these sites affected the peak pattern evenness obtained with different dye sets as shown in Table 4.1.

Although we have focused mainly on *Taq* Pol I, we also have studied polymerases derived from *Thermatoga maritima* and *Thermus thermophilus*. In addition to looking at differences in peak pattern evenness with these enzymes, we have examined other properties including salt tolerance and the ability to read through certain G-rich regions. Enzymes that are salt

Chapter 4. New DNA Sequencing Enzymes

Table 4.1. Effect of the enzyme on peak pattern evenness in sequencing reactions with different dye terminators.

Enzyme	Average relative error determined for each enzyme normalized to that obtained with AmpliTaq DNA Polymerase FS (7)			
	Rhodamine Dye Terminators[a]	d-Rhodamine Dye Terminators[b]	BigDye Terminators[b]	BigDye Terminators v3.0[b]
1	0.38	0.74	1.04	1.07
2	NA	1.03	1.22	1.1
3	0.64	1.00	1.20	NA
4	0.62	0.82	1.39	1.07
5	0.95	1.36	1.46	1.35
6	0.73	1.25	1.17	1.06
7	1	1	1	1

[a] 377 data.
[b] 3100 data.

Sequence data obtained with the different enzymes was analyzed to determine the relative error for each base reaction (see **Methods and Materials** for a definition of relative error). The average relative error of the four base reactions was then determined. The average relative error measured for the different enzymes was normalized to that obtained for data generated with AmpliTaq DNA Polymerase FS. Numbers less than 1 indicate data with more even peak heights for the four base reactions overall. Enzymes are: (1) modified *Tma* Pol I; (2) modified *Tth* Pol I; (3) E681H; (4) T664G; (5) R660I; (6) R660P; and (7) AmpliTaq DNA Polymerase FS.

tolerant also have shown more resistance to contaminants in sample preparation (1, 3, 52). In certain G-rich regions, *Taq* Pol I stops or there is a significant reduction of signal after the region. We have found interesting differences in the polymerases derived from *T. maritima* and *T. thermophilus* when compared to *Taq* Pol I.

Methods and Materials

Enzymes

Modified enzymes from *T. maritima* and *T. thermophilus* were gifts from Dr. David Gelfand (Roche Molecular Systems). Because native Pol I-type polymerases from *Thermotoga* genera have both 5'-3' structure specific nuclease and 3'-5' exonuclease activities, both these activities need to be removed for the enzyme to be used in sequencing reactions. A mutation equivalent to *Taq* F667Y was introduced as well. The variants of *Taq* Pol I, which have F667Y and G46D, were developed in collaboration with Dr. Jack Richards and Dr. Curtis Bloom (8). The variants used in this study were E681H, T664G, R660I, and R660P.

Sequencing Reactions

Reaction premixes were prepared using buffer, dNTPs, enzyme and the dye terminator set of interest. Twenty microliter reactions were prepared with 8 µl of premix, 3.2 pmoles of primer, 0.25 to 0.5 µg of double stranded template and other additions as noted. The standard cycling program used was 96°C for 10 seconds, 50°C for 5 seconds, and 60°C for four minutes for 25 cycles. After cycling the unincorporated terminators were removed using spin columns (Princeton Separation) and the reactions were dried. Samples were resuspended in formamide mixed with Blue Dextran loading solution for analysis on the 377 or formamide alone for the 3100. Reaction products were analyzed on either an ABI Prism 377 DNA Sequencer or an ABI Prism 3100 Genetic Analyzer.

Read Length

The read length obtained in sequencing reactions was estimated based on the number of bases having a quality value, q, equal to or greater than 20, which is determined using Phred (19). A q value of 20 indicates that the probability that the base call is incorrect is 1/100 (1%). A q value of 30 indicates that the probability that the base call is incorrect is 1/1000 (0.1%).

Peak Height Evenness

Peak height evenness was measured as relative error. The relative error is the ratio of the standard deviation to the mean peak height, where the mean peak height is the average peak height of all acceptable peaks and the standard deviation is the square root of the average squared deviation of acceptable peak heights from their mean. A lower relative error indicates more even peak heights.

Results and Discussion

Evaluation of Peak Pattern Evenness and Incorporation of Labeled Dideoxynucleotides with Different Enzymes

The peak pattern evenness in sequence data obtained with AmpliTaq DNA Polymerase FS and d-rhodamine, BigDye Terminators or BigDye Terminators v 3.0 is much better than was obtained with the original rhodamine dye terminators. However, the peak pattern evenness still may

Chapter 4. New DNA Sequencing Enzymes

be a limiting factor when trying to detect low levels of mutation (10% to 20%) at a particular peak position. Because the amplitude variation of peaks is dependent on the sequence context and particular base involved, the reliability of mutation detection varies depending on which bases are involved. Minimizing these sequence context effects and getting very even peak heights is important in improving the reliability of detecting mixed bases. Table 4.1 shows the results obtained for peak pattern evenness with different enzymes and different dye sets. The ability to affect peak pattern evenness varies with the different dye sets as well as the different enzymes. The most variation can be seen with the rhodamine dye terminators, which were used in the original dye terminator chemistry. The biggest improvement in peak pattern evenness with this dye set was obtained with modified *Tma* Pol I. Significant improvement also was seen with E681H and T664G. Other mutants at 681 were tested as well and 13 out of 16 variants tested showed improvement in the peak pattern evenness with rhodamine dye terminators (unpublished data). Similar results were obtained with the d-rhodamine dye terminators but the magnitude of the effects was reduced. In contrast to the results obtained with mutants at 681 and 664, only R660I showed significant improvement with rhodamine dyes. Both 660 variants have less even peak patterns with d-rhodamine dye terminators. The rhodamine and drhodamine dye terminators have a single dye label and generally similar structure (4, 54). With the BigDye terminators and BigDye Terminators v 3.0 the ability to impact the peak pattern was reduced. Modified *Tma* Pol I or E681H or T664G did not produce data with significantly more even peak heights than AmpliTaq DNA Polymerase FS. With all four dye sets, the peak pattern evenness in sequence data obtained with modified *Tth* Pol I was very similar to the results obtained with AmpliTaq DNA Polymerase FS. Although the two O-helix mutants at 660 that were tested here show less even peak patterns, they do show significant changes in the distribution of signal in the four base reactions with BigDye Terminators (Table 4.2). For R660I 75% of the total signal was in the C reaction as compared to only 30.5% for AmpliTaq DNA Polymerase FS with the same pool of terminators. With R660P the increase was not as dramatic, but the signal in C was still increased to 44% of the total signal. In fact all 18 variants tested at this site showed more than a 10% increase in the signal in the C reaction with BigDye Terminators (unpublished data). Such changes were less noticeable with the BigDye Terminators v3.0, but R660I showed a significant increase in the distribution of signal in G (Table 4.2).

Salt Tolerance

Enzymes also differ in their ability to tolerate salts such as KCl and other substances (1). This has been particularly important for PCR because car-

Table 4.2. Distribution of signal in the four base reactions for *Taq* variants with BigDye Terminators and BigDye Terminators V3.0.

Enzyme	Total Signal, %			
	G	A	T	C
BigDye Terminators				
E681H	9	39	23	29
T664G	10	27	25	38
R660I	6	4	15	75
R660P	10	11	31	48
AmpliTaq DNA Polymerase FS	13	30	27	30
BigDye Terminators V3.0				
E681H	36	24	19	21
T664G	39	24	16	20
R660I	63	21	6	10
R660P	43	17	19	21
AmpliTaq DNA Polymerase FS	46	21	15	18

In this experiment a common pool of terminators was used for sequencing reactions with AmpliTaq DNA Polymerase FS and the *Taq* variants. The numbers reported are the % of total signal in each of the four base reactions.

ryover contamination from genomic DNA samples can be significant. Of the enzymes tested here only modified *Tth* Pol I shows significant salt tolerance (Table 4.3). The variants of *Taq* Pol I that were tested showed no difference from AmpliTaq DNA Polymerase FS in the ability of the enzyme to sequence in the presence of KCl. The ability of *Tth* DNA Polymerase to tolerate salt is consistent with other reports in the literature of the ability of this enzyme to tolerate salts as well as other substances (1, 3, 52).

Ability to Read Through Difficult Regions

To produce a reliable sequencing chemistry the DNA polymerase needs to be able to sequence through a wide range of sequence contexts. At some of these sites there may be regions of secondary structure that affect the ability of the enzyme to read through the region. As a component of the

Chapter 4. New DNA Sequencing Enzymes

Table 4.3. Salt tolerance.

Enzyme	Total Signal		% of Control without KCl	Number of Bases with $q = 20$ (Phred)		
	Control	100 mM KCl		Control	100 mM KCl	Change
1	5054	258	5.1	619	253	−366
2	4353	1144	26.0	583	574	−9
3	5949	143	2.8	582	81	−501
4	4723	227	4.8	577	213	−364
5	3011	169	5.6	536	158	−378
6	4025	250	6.2	584	199	−385
7	4263	268	6.2	552	166	−386

Sequencing reactions were generated with a moderately GC-rich template either without KCl (control) in the reaction or with 100 mM KCl. The total signal in the four base reactions is shown for each enzyme tested as well as the % of signal remaining in the reactions carried out in the presence of 100 mM salt. The number of bases was measured as the number of bases with a quality value, q, equal to 20 or greater as determined using Phred (19) (see **Methods and Materials**). Enzymes are: 1) modified *Tma* Pol I; 2) modified *Tth* Pol I; 3) E681H; 4) T664G; 5) R660I; 6) R660P; and 7) AmpliTaq DNA Polymerase FS.

standard reaction, dITP is commonly used and affects the duplex stability of the extending strand. This also may result in stops in very G-rich regions. Plasmid 2358 is a clone derived from human chromosome 19. This template has a relatively G-rich region followed by a region in which a hairpin with a ΔG of −8.6 kcal/mole can form. Each of the enzymes tested here has been evaluated for its ability to sequence through this region (Table 4.4). The average peak height of 10 bases (bases 209 to 218) after the stop region and ten bases (bases 179–188) before this region was determined. The ratio of these two numbers indicates the efficiency with which the enzyme has read through the region. For AmpliTaq DNA Polymerase FS this number is 0.17, which is consistent with the significant reduction of signal in this region as shown in Figure 4.1B. The variants of *Taq* Pol I that were tested here also showed a significant decrease in signal after this region with ratios from 0.1 to 0.18, which were comparable to the numbers obtained with AmpliTaq DNA Polymerase FS. This correlated with a reduction in the number of bases with Phred Q20. However, the enzyme from *Thermus thermophilus* showed distinctly different behavior and typically read through this region with a much smaller reduction in signal (Table 4.4 and Fig. 1A). The results with modified *Tma* Pol I was similar to that observed with modified *Tth* Pol I.

Table 4.4. Ability of different enzymes to read through a stop region.

Enzyme	A	B	Ratio (B/A)	Number of Bases with $q = 20$ (Phred)
1	460	365	0.79	646
2	778	510	0.65	624
3	458	46	0.10	169
4	774	183	0.24	520
5	679	140	0.21	440
6	862	154	0.18	491
7	756	132	0.17	390

Sequence:
179 GCCAGGGGTA GGGGTAGGGG TTGGGGTGGC AGGGCCCTGG 219 TAACCCTGCT CCCCGGCCCC CCAGATCCTG CATGCCAATG The underlined sequence can form a hairpin with ΔG of −8.6 kcal/mole.
Template p2358 has a G-rich region followed by a region of secondary structure 206 bases from the primer. The sequence of this region is shown below. In this case the signal was determined by measurement of the peak height in the raw data of 10 bases before this region (bases 179–188) and 10 bases after this region (bases 209–218). The average signal for 179–188 (A) and 209–218 (B) was determined for each enzyme. The number of bases was measured as the number of bases with a quality value, q, equal to 20 or greater using Phred (19) (see **Methods and Materials**). With AmpliTaq DNA Polymerase FS there is a significant decrease in the signal just before the sequence GCAGGG. Reactions were analyzed on an ABI Prism 3100 DNA Analyzer with POP-6. Enzymes are: 1) modified *Tma* Pol I; 2) modified *Tth* Pol I; 3) E681H; 4) T664G; 5) R660I; 6) R660P; and 7) AmpliTaq DNA Polymerase FS.

Conclusion

In the work described here we have looked at enzymes derived from three different sources. *Tma* Pol I is from *Thermatoga maritima*, a different genus of eubacteria. *Tth* Pol I is from a different species of *Thermus*, *Thermus thermophilus*. In addition, we have tested variants of *Taq* Pol I with mutations at two different sites in the O helix and at an additional site in the adjacent fingers region. Significant differences are seen in the results obtained with these enzymes. Modified *Tma* DNA polymerase shows more even peak heights with both rhodamine and drhodamine dye terminators than does either AmpliTaq DNA Polymerase FS or modified *Tth* DNA Polymerase. Two groups reported the use of the DNA Polymerase from the related organism *Thermatoga neapolitana* for DNA sequencing (5, 58). However, the peak height evenness with BigDye Terminators and with BigDye Terminators v 3.0 is not significantly better. In contrast to the results obtained with *Tma* Pol I, the ability of *Tth* Pol I to incorporate the

Chapter 4. New DNA Sequencing Enzymes

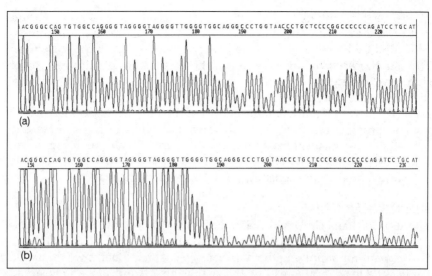

Figure 4.1. Comparison of data obtained with modified *Tth* Pol I and AmpliTaq DNA polymerase FS in the region of a hairpin. The template used is that described in Table 4.4. Electropherograms obtained with modified *Tth* Pol I (a) and AmpliTaq DNA Polymerase FS (b) using BigDye Terminators V3.0. Reactions were analyzed on an ABI Prism 3100 Genetic Analyzer.

various dye terminator sets used here is very similar to that of AmpliTaq DNA Polymerase FS. However, this enzyme does show significant differences in other properties such as salt tolerance and the ability to read through a stop region. In *de novo* sequencing, the ability to sequence a higher proportion of templates successfully is an advantage because it increases the reliability of the process.

Variants of *Taq* Pol I that have mutations in the O-helix and adjacent fingers region have been shown to affect peak pattern evenness and the ability to incorporate different bases. The results vary with the particular dye set involved and the amino acid substitution at a particular site. The ability of mutations in this region to affect incorporation of different bases is consistent with the involvement of the O-helix in nucleotide binding. However, the molecular models show that the linker arms and dye atoms are located well away from the O-helix. In other properties the *Taq* Pol I variants behave like AmpliTaq DNA Polymerase FS. For example, these mutations had no impact on the salt tolerance of the enzyme or the ability of the enzyme to sequence through a region of secondary structure.

Clearly, the availability of better DNA polymerases can play an important role in achieving improved data quality as well as improving reliability. In the last few years several crystal structures of *Taq* Pol I, both in the open and closed conformation, have been published (10, 39, 40, 41).

In conjunction with extensive mutagenesis we now have a much clearer picture of the enzyme structure and function (51). Comparative alignment of *Taq* Pol I with sequences of the closely related DNA Polymerase from *Thermus thermophilus* can give us additional tools to identify regions of significant differences. By correlating changes in the polymerase with differences in the behavior of the DNA Polymerases in DNA sequencing, we can learn how to manipulate properties of the polymerase to improve its general properties as well as specifically its ability to incorporate dye labeled dideoxynucleotides. In the future we may be able to design novel DNA polymerases for specific applications that will lead to improvements in both the quality and reliability of DNA sequencing reactions.

Acknowledgments

The authors thank Dr. David Gelfand of Roche Molecular Systems for supplying the modified *Tma* Pol I and the modified *Tth* Pol I. We also thank Gil Amparo for supplying the data for this chapter. Finally, we thank Dr. Paolo Vatta for critically reading this manuscript. AmpliTaq is a registered trademark of Roche Molecular Systems.

References

1. Al-Soud, W.A., and Rådström, P. 1998. Capacity of nine thermostable DNA polymerases to mediate DNA amplification in the presence of PCR-inhibiting samples. *Appl Environ Microbiol* 64: 3748–3753.
2. *Automated DNA Sequencing Chemistry Guide.* 1998. Applied Biosystems. Foster City, CA.
3. Bélec, L.J., Authier, M-C., Eliezer-Vanerot, C., et al. 1998. Myoglobin as a polymerase chain reaction (PCR) inhibitor: a limitation for PCR from skeletal muscle tissue avoided by the use of *Thermus thermophilus* polymerase. *Muscle Nerve* 21: 1064–1067.
4. Bergot, B.J., Chakerian, V., Connell, C.R., et al. 1989. Spectrally resolvable rhodamine dyes for nucleic acid sequence determination. U.S. Patent 5,366,860, November 22, 1994.
5. Blakesley, R.W., Xu, L., Westfall, B., et al. 1996. High performance DNA sequencing data using *Tne* DNA polymerase. *Microb Comp Genomics* 1: 225.
6. Braithwaite, D.K., and Ito, J. 1993. Compilation, alignment, and phylogenetic relationships of DNA polymerases. *Nucleic Acids Res* 21: 787–802.
7. Brandis, J.W. 1999. Dye structure affects Taq DNA polymerase terminator selectivity. *Nucleic Acids Res* 27: 1912–1918.
8. Brandis, J., Bloom, C., and Richards, J.H. 2001. DNA polymerases having improved labeled nucleotide incorporation properties. U.S. Patent 6,265,193.
9. Brandis, J.W., Edwards, S.G., and Johnson, K.A. 1996. Slow rate of phosphodiester bond formation accounts for the strong bias that *Taq* DNA polymerase shows against 2', 3'-dideoxynucleotide terminators. *Biochemistry* 35: 2189–2200.

10. Brautigam, C.A., and Steitz, T.A. 1998. Structural and functional insights provided by crystal structures of DNA polymerases and their substrate complexes. *Curr Opin Struct Biol* 8: 54–63.
11. Carothers, A.M., Urlaub, G., Mucha, J., et al. 1989. Point mutation analysis in a mammalian gene: rapid preparation of total RNA, PCR amplification of cDNA, and *Taq* sequencing by a novel method. *BioTechniques* 7: 494–499.
12. Chen, S-M., Amparo, G. Kuo, E., et al. 2001. Improvements in ABI Prism dGTP BigDye Terminator V3.0 Ready Reaction Cycle Sequencing Chemistry. *Program and Abstract Journal GSAC* 13: 58.
13. Chung, A.P., Rainey, F.A., Valente, M., et al. 2000. *Thermus igniterrae* sp. Nov. and *Thermus antranikianii* sp. Nov., two new species from Iceland. *Int J Syst Evol Microbiol* 50: 209–217.
14. Connell, C., Fung, S., Heiner, C., et al. 1987. Automated DNA sequence analysis. *BioTechniques* 5: 342–348.
15. Craxton, M. 1991. Linear amplification sequencing, a powerful method for sequencing DNA. *Methods: A Companion to Methods in Enzymology* 3: 20–26.
16. Dicker, A.P., Volkenandt, M., Adamo, A., et al. 1989. Sequence analysis of a human gene responsible for drug resistance: a rapid method for manual and automated direct sequencing of products generated by the polymerase chain reaction. *BioTechniques* 7: 830–837.
17. Eun, H-M. 1996. *Enzymology Primer for Recombinant DNA Technology*. Academic Press. San Diego.
18. Evans, S.J., Fogg, M.J., Mamone, A., et al. 2000. Improving dideoxynucleotide-triphosphate utilisation by the hyper-thermophilic DNA polymerase from the archaeon *Pyrococcus furiosius*. *Nucleic Acids Res* 28: 1059–1066.
19. Ewing, B., Hillier, L., Wendl, M.C., and Green, P. 1998. Base-calling of automated sequencer traces using Phred. II. Error probabilities. *Genome Res* 8: 186–194.
20. Gardner, A.F., and Jack, W.E. 1999. Determinants of nucleotide sugar recognition in an archaeon DNA polymerase. *Nucleic Acids Res* 27: 2545–2553.
21. Gardner, A.F., and Jack, W.E. 2002. Acyclic and dideoxy terminator preferences denote divergent sugar recognition by archaeon and *Taq* DNA polymerases. *Nucleic Acids Res* 30: 605–613.
22. Gelfand, D.H., Lawyer, F.C., and Stoffel, S. 1995. Mutated thermostable nucleic acid polymerase enzyme from *Thermotoga maritima*. U.S. Patent 5,420,029.
23. Gocayne, J., Robinson, D.A., FitzGerald, M.G., et al. 1987. Primary structure of rat cardiac B-adrenergic and muscarinic cholinergic receptors obtained by automated DNA sequence analysis: further evidence for a multigene family. *Proc Natl Acad Sci USA* 84: 8296–8300.
24. Goodman, M.F. 2002. Error-prone repair DNA polymerases in prokaryotes and eukaryotes. *Annu Rev Biochem* 71: 17–50.
25. Heiner, C., and Hunkapiller, T. 1989. Automated DNA sequencing. In Howe, C.J., and Ward E.S., eds. *Nucleic Acids Sequencing*. IRL Press at Oxford University Press. Oxford, UK.
26. Hughes, A.J. Jr., and Chatterjee, D.K. 1999. Cloned DNA polymerases from *Thermotoga neapolitana* and mutants there of. U.S. Patent 5,939,301. August 17, 1999.

27. Inman, R.B., and Baldwin, R.L. 1964. Helix-random coil transitions in DNA homopolymer pairs. *J Mol Biol* 8: 452–469.
28. Innis, M.A., and Gelfand, D.H. 1999. Optimization of PCR: Conversations between Michael and David. In Innis, M.A., Gelfand, D.H., and Sninsky, J.J., eds. *PCR Applications: Protocols for Functional Genomics*. Academic Press. San Diego.
29. Innis, M.A., Myambo, K.B., Gelfand, D.H., and Brow M.A.D. 1988. DNA sequencing with *Thermus aquaticus* DNA polymerase and direct sequencing of polymerase chain reaction-amplified DNA. *Proc Natl Acad Sci USA* 85: 9436–9440.
30. International Human Genome Sequencing Consortium. 2001. Initial sequencing and analysis of the human genome. *Nature* 409: 860–921.
31. Ito, J., and Braitwaite, D.K. 1991. Compilation and alignment of DNA polymerase sequences. *Nucleic Acids Res* 19: 4045–4057.
32. Johnston-Dow, L., Mardis, E., Heiner, C., and Roe, B.A. 1987. Optimized methods for fluorescent and radio-labeled DNA sequencing. *BioTechniques* 5: 754–765.
33. Kalman, L.V. 1995. Thermostable DNA polymerases with altered discrimination properties. *Genome Sci Technol* 1: P-42 (abstract).
34. Khan, S. H., Rosenblum, B.B., Zhen, W., et al. 1999. Dye terminator chemistry: effects of substrate structure on sequencing analysis. *Proceedings of Advances in Fluorescence Sensing Technology IV* 3602: 367–378.
35. Kong, H., Kucera, R.B., and Jack, W.E. 1993. Characterization of a DNA polymerase from the hyperthermophile archaea *Thermococcus litoralis*. *J Biol Chem* 268: 1965–1975.
36. Kornberg, A., and Baker, T. 1992. *DNA Replication*, 2nd ed. Freeman: New York.
37. Lawyer, F.C., Stoffel, S., Saiki, R.K., et al. 1993. High level expression, purification and enzymatic characterization of full-length *Thermus aquaticus* DNA polymerase and a truncated form deficient in 5'-3' exonuclease activity. *PCR Methods Appl* 2: 275–287.
38. Lee, L.G., Spurgeon, W.L., Heiner, C.R., et al. New energy transfer dyes for DNA sequencing. *Nucleic Acids Res* 25: 2816–2822.
39. Li, Y., and Waksman, G. 2001. Crystal structures of a ddATP-, ddTTP-, ddCTP-, and ddGTP-trapped ternary complex of Klentaq1: insights into nucleotide incorporation and selectivity. *Protein Sci* 10: 1225–1233.
40. Li, Y., Korolev, S., and G. Waksman, G. 1998. Crystal structures of open and closed forms of binary and ternary complexes of the large fragment of *Thermus aquaticus* DNA polymerase I: structural basis for nucleotide incorporation. *EMBO J* 17: 7514–7525.
41. Li, Y., Kong, Y., Korolev, S., and Waksman, G. 1998. Crystal structures of the Klenow fragment of *Thermus aquaticus* DNA polymerase I complexed with deoxyribonucleoside triphosphates. *Protein Sci* 7: 1116–1123.
42. Manaia, C.M., Hoste, B., Gutierrez, M.C., et al. 1994. Halotolerant *Thermus* strains from marine and terrestrial hot springs belong to *Thermus thermophilus* (ex Oshima and Imahori, 1974) nom. rev. emend. *System Appl Microbiol* 17: 526–532.
43. Meldrum, D. 2000. Automation for genomics, part one: preparation for sequencing. *Genome Res* 10: 1081–1092.

Chapter 4. New DNA Sequencing Enzymes 53

44. Meldrum, D. 2000. Automation for genomics, part two: sequencers, microarrays, and future trends. *Genome Res* 10: 1288–1303.
45. Mills, D.R., and Kramer, F.R. 1979. Structure-independent nucleotide sequence analysis. *Proc Natl Acad Sci USA* 76: 2232–2235.
46. Moreira, L.M., Da Costa, M.S., Sa-Correia, I. 1997. Comparative genomic analysis of isolates belonging to the six species of the genus *Thermus* using pulsed-field gel electrophoresis and ribotyping. *Arch Microbiol* 168: 92–101.
47. Murray, V. 1989. Improved double-stranded DNA sequencing using the linear polymerase chain reaction. *Nucleic Acids Res* 17: 8889.
48. Parker, L.T., Deng, Q., Zakeri, H., et al. 1995. Peak height variations in automated sequencing of PCR products using *Taq* Dye-Terminator Chemistry. *BioTechniques* 19: 116–121.
49. Parker, L.T., Zakeri, H., Deng, Q., et al. 1996. AmpliTaq DNA Polymerase, FS Dye-Terminator sequencing: analysis of peak height patterns. *BioTechniques* 21: 694–699.
50. Patel, P., and Loeb, L.A. 2000. DNA polymerase active site is highly mutable: evolutionary consequences. *Proc Natl Acad Sci USA* 97: 5095–5100.
51. Patel, P.H., Suzuki, M., Adman, E., et al. 2001. Prokaryotic DNA Polymerase I: evolution, structure, and "base flipping" mechanism for nucleotide selection. *J Mol Biol* 308: 823–837.
52. Poddar, S.K., Sawyer, M.H., and Connor, J.D. 1998. Effect of inhibitors in clinical specimens on Taq and Tth DNA polymerase-based PCR amplification of influenza A virus. *J Med Microbiol* 47: 1131–1135.
53. Prober, J.M., Trainor, G.L., Dam, R.J., et al. 1987. A System for rapid DNA sequencing with fluorescent chain-terminating dideeoxynucleotides. *Science* 238: 336–341.
54. Rosenblum, B.B., Lee, L.G., Spurgeon, S., et al. 1997. New dye-labeled terminators for improved DNA sequencing patterns. *Nucleic Acids Res* 25: 4500–4504.
55. Saiki, R.K., Gelfand, D.N., Stoffel, S., et al. 1988. Primer-directed enzymatic amplification of DNA with a thermostable DNA polymerase. *Science* 239: 487–491.
56. Sanger, F., Nicklen, S., and Coulson, A.R. 1977. DNA sequencing with chain-terminating inhibitors. *Proc Natl Acad Sci USA* 74: 5463–5467.
57. Sears, L.E., Moran, L.S., Kissinger, C., et al. 1992. CircumVent™ Thermal Cycle Sequencing and alternative manual and automated sequencing protocols using the highly thermostable Vent$_R$™ (exo⁻) DNA Polymerase. *BioTechniques* 13: 626–633.
58. Slater, M.R., Hartnett, J.R., Huang, F., et al. 1995. DNA Polymerase I of *Thermotoga neapolitana (Tne)* and mutant derivatives. *Genome Sci Technol* 1: P-47.
59. Slater, M.R., Huang, F., Hartnett, J.R., et al. 2000. Thermophilic DNA polymerases from *Thermotoga neapolitana*. U.S. Patent 6,077,664, June 20, 2000.
60. Spurgeon, S., Koepf, S., Heiner, C., and Abramson R. 1994. Improved Taq Dye Primer cycle sequencing using AmpliTaq DNA Polymerse CS. Genome Sequencing and Analysis Conference VI. pA-30 (abstract).
61. Tabor, S., and Richardson, C.C. 1987. DNA sequence analysis with a modified bacteriophage T7 DNA polymerase. *Proc Natl Acad Sci USA* 84: 4767–4771.

62. Tabor, S., and Richardson, C.C. 1990. DNA sequence analysis with a modified bacteriophage T7 DNA Polymerase: effect of pyrophosphorolysis and metal ions. *J Biol Chem* 265: 8322–8328.
63. Tabor, S., and Richardson, C.C. 1995. A single residue in DNA polymerases of the *Escherichia coli* DNA polymerase I family is critical for distinguishing between deoxy and dideoxyribonucleotides. *Proc Natl Acad Sci USA* 92: 6339–6343.
64. Venter, J.C., Adams, M.D., Myers, E.W., et al. 2001. The sequence of the human genome. Science 291: 1304–1351.
65. Williams, R.A.D., Smith, K.E., Welch, S.G., et al. 1995. DNA relatedness of *Thermus* strains, description of *Thermus brokianus* sp. Nov., and a proposal to reestablish *Thermus thermophilus* (Oshima and Imahori). *Int J Syst Bacteriol* 45: 495–499.
66. Zakeri, H., Amparo, G., Chen, S.-M., et al. 1998. Peak height pattern in dichloro-rhodamine and energy transfer dye terminator sequencing. *BioTechniques* 25: 406–413.

5 Beyond pUC: Vectors for Cloning Unstable DNA

Ronald Godiska, Melodee Patterson,
Tom Schoenfeld, and David A. Mead
Lucigen Corporation, Middleton, WI

Introduction

The foundation of genomic sequence analysis is large-scale cloning and sequencing from shotgun plasmid libraries. Phenomenal advances in the technology for obtaining and assembling sequence information from vast numbers of clones have made this approach practical and efficient. In stark contrast, there has been very little improvement in the vectors used for generating plasmid libraries for shotgun sequencing. Common vectors are typically maintained at high copy number and induce transcription and translation of inserted fragments, causing instability of certain classes of DNA sequences. This "unstable" DNA may result in sequence stacking, clone gaps, or severe difficulties in creating plasmid libraries, especially from DNA with a high percentage (>70%) of adenine and thymine bases (AT-rich DNA).

By far the most commonly used plasmid for constructing shotgun libraries is pUC18 or its closely related derivatives. These vectors have several features widely accepted as advantageous, such as their blue/white screening capacity, large multiple cloning site, and high copy number, as well as the ability to generate RNA transcripts from bacteriophage promoters and single-stranded DNA (ssDNA) from the M13 origin of replication (reviewed in 37). However, very little attention has been devoted to characterizing the potential disadvantages of these traits. For example, specific fragments or even large portions of genomes are recalcitrant to cloning in these vectors, hindering most sequencing efforts. As a result, despite extensive finishing efforts applied over the past years, gaps remain in the genomic sequence of nearly every multicellular organism studied, including those assemblies described as "complete" (36).

The following review summarizes difficulties associated with cloning into the multipurpose pUC-based plasmids. It also presents data showing that dramatic improvements in library construction are possible with new vectors that reduce the cloning bias. For example, telomeric repeats and other AT-rich fragments from *Pneumocystis carinii* (60% to 65% AT) were stable in a low copy, transcription-free vector but were unstable in pUC19. Likewise, the AT-rich genome of *Lactobacillus helveticus* (65% AT) was cloned with 25-fold greater efficiency and significantly less bias using a transcription-free vector. Toxic regions of the mouse hepatitis virus genome were readily cloned and stable in the new type of vector, but they were deleted, rearranged, or slow-growing in conventional vectors. Although successful genomic sequencing with pUC-based vectors has validated their utility and reinforces their continued use, employing alternative vectors may substantially reduce the effort required for genomic sequencing and assembly.

Drawbacks of pUC Vectors

Gaps in shotgun libraries and seemingly "unclonable" DNA fragments are quite common. Such DNA is typically characterized by high AT-content, strong secondary structure, open reading frames, or *cis*-acting functions (e.g., transcriptional promoters or replication origins). In some cases, most notably AT-rich DNA, the reasons for difficulty in cloning are not well defined. In other instances, one or more features of pUC plasmids have been shown to be incompatible with cloning or stable maintenance of the inserts. Difficulties caused by each of these features are summarized in Table 5.1 and described below in detail.

Vector-Driven Transcription

The central feature of pUC plasmids, which has become ubiquitous among cloning vectors, is the "blue/white" colony screen to detect recombinant plasmids (61). This screen is based upon inactivation of the *lac*Zα peptide of beta-galactosidase, which is expressed by the vector. A similar approach is used with direct selection vectors (e.g., pZeRO; Invitrogen, Carlsbad, CA), except that a gene encoding a lethal product, such as *ccd*B or *sac*B, is used in place of or in addition to *lac*Zα to select against non-recombinants.

Although these screens provide a simple and powerful method to identify most recombinant colonies, they may in fact select against a relatively large number of clones. Both screens induce a high level of transcription and translation of the indicator gene, driven by a promoter in the vector region adjacent to the cloning site. Sequences inserted into the

Chapter 5. Beyond pUC: Vectors for Cloning Unstable DNA

Table 5.1. Advantages and disadvantages of pUC-type plasmids.

Feature	Advantage	Disadvantage	Targets Selected Against
Blue/white screen or direct selection	Easy screening	Transcription/translation of inserts, false positives, false negatives, or loss of recombinant clones	Toxic ORFs, repeats, AT-rich DNA, short inserts, promoters
High copy number	High plasmid yield	Instability	Large inserts, AT-rich DNA
Ampicillin resistance	Common, inexpensive	Satellite colonies, loss of selection in liquid culture	Slow growers, unstable inserts
M13 origin	Generates ssDNA	Instability when transcribed	Promoters, cDNAs

multiple cloning site are also subject to this transcription and translation, which may interfere with cloning of several classes of inserts, as described below.

Open Reading Frames (ORFs)

Fragments encoding enzymatic activities or structural proteins are often difficult to clone, due to expression of their product from the vector-borne promoter. Methods to avoid lethality in such cases include engineering the product for secretion (48), cloning the gene in several segments (69), or directional cloning of the ORF in the "reverse" orientation relative to transcription from the vector's promoter (62). Successful cloning of the inserts can be verified by direct sequence analysis or by biochemical assays of the encoded product.

Because such approaches are not practical for library construction, toxic coding sequences are likely to be underrepresented or not present at all. Loss of clones encoding deleterious peptides usually goes unnoticed until gaps in the assembly are detected. The unintended selection may be particularly important when cloning viral or cDNA libraries, because a wide variety of coding sequences may hinder cell growth if expressed at high levels [e.g., viral peptides (26, 34, 45, 66), Pfu DNA polymerase (14), pectate lyase (58), cyclomaltodextrinase (32), etc.].

Problems due to overexpression of proteins may be lessened by repressing the vector's promoter, which in turn inactivates the blue/white screen or direct selection. For example, at least partial repression of the *lacZ* promoter in pUC-type plasmids may be achieved by using a vector or a bacterial strain that harbors the *lac*Iq repressor allele, which minimizes transcription from the *lacZ* promoter (62). Vectors that carry *lac*Iq include pET (Novagen, Madison, WI), pGEX (Promega, Madison, WI), and pMAL-c2 (New England Biolabs, Beverly, MA). An F' plasmid that carries the *lac*Iq gene also is present in many bacterial strains, including JM101, DH5 (Invitrogen), E. cloni 10GF' (Lucigen, Madison, WI), and XL-1 Blue (Stratagene, La Jolla, CA). The caveat to inactivating the screen is that the background of empty vector clones is indistinguishable from the recombinants; therefore, it must be reduced to an acceptable level. For library construction, the simplest method of reducing background is dephosphorylation of the vector. Alternatively, vectors can be designed with noncompatible, single-stranded extensions, such as those produced by appropriately designed BstXI restriction sites. The fragments to be cloned are first ligated to oligonucleotide linkers containing extensions that are not self-complimentary but are complimentary to the ends of the vector. They are subsequently ligated to the vector. The noncomplimentary ends prevent self-ligation of the vector and concatenation of insert fragments.

Repeats

Due to their genetic instability in the genomes of eukaryotes and prokaryotes, tracts of trinucleotide or dinucleotide repeats have attracted much interest. In numerous human diseases, such as Huntington's disease, fragile X syndrome, and myotonic dystrophy (reviewed in 3 and 67), the repeats at a particular locus are deleted or expanded by tens or even thousands of copies. The resulting sequence may alter the product or expression of an encoded transcript, or it may be prone to DNA breakage.

The mechanisms responsible for expanding or contracting the repeats in mammalian cells appear to operate in *E. coli* as well. Genomic repeats and defined synthetic repeat sequences, therefore, can be cloned into plasmids and analyzed *in vitro* to characterize *cis*- and *trans*-acting elements (5, 8, 9, 22, 24, 44). However, the ability of bacteria to rearrange these sequences is likely to interfere with obtaining authentic genomic clones, so minimizing this process is essential for obtaining accurate genomic sequence information.

Bacterial studies have demonstrated that the sequence of the repeat as well as the mode and level of plasmid replication and transcription may play a role in expansion and deletion of repeats. When cloned into pUC19, the CTG repeat is expanded at least eightfold more frequently

Chapter 5. Beyond pUC: Vectors for Cloning Unstable DNA

than nine other sequences tested (41). Deletion and expansion of repeat segments has been suggested to occur via the formation of DNA hairpins during repair (44) or during replication, on either the leading (24) or the lagging strand (52) of the replication fork.

The frequency of deletion of cloned repeats increases up to 20-fold upon induction of the *lacZ* promoter, which drives transcription of inserts in pUC19 (8). The deleted plasmids appear to have a strong growth advantage upon subculturing, obscuring the original nondeleted molecules (8, 9, 58). Minimizing plasmid transcription is, therefore, critical to maintain the integrity of cloned trinucleotide repeats.

Dinucleotide repeats are also unstable in *E. coli*. The main factor affecting their stability appears to be the formation of secondary structure during replication (5). Interestingly, induction of transcription does not appear to affect the stability of dinucleotide repeats cloned into pUC19, even under conditions that induce instability of trinucleotide repeats.

Replication

Transcription also has been reported to induce deletions of inserts from plasmids containing an active M13 origin of replication (60, 64, 65). These studies suggest that collision of the replication and transcription complexes leads to frequent deletions, which originate at a nick at the M13 origin and extend downstream into the transcribed region. Deletion is dependent on nicking of the M13 origin by the phage replication protein II (gpII) and occurs whether replication is mediated via the pBR322 origin or the M13 origin (64).

False Positives and False Negatives

A significant problem with the blue/white screen is the presence of false negative and false-positive colonies, as well as results that cannot be easily classified (55). False positives are colonies that appear white but do not contain a foreign DNA insert in the *lacZα* cloning site. Factors responsible for generating false positives are: (1) exonuclease contamination that removes bases from the termini of the *lacZα* region during vector processing, creating frame-shifts that inactivate the peptide; (2) spontaneous mutations in the gene for the *lacZα* fragment or the *lacZ*Δ*M15* allele; and (3) loss of the F' episome carrying the *lacZ*Δ*M15* allele.

False-positive colonies are carried forward and analyzed as authentic positive clones, wasting time and resources. A further problem with false positives is that they may obscure the presence of true positives. This issue is most significant when very few recombinants are expected, such as in attempting to clone fragments that are large (e.g., >10Kb), in low amounts (e.g., less than 5–10 ng), or difficult to clone for various other reasons.

False negatives are clones that do in fact contain DNA inserted into the cloning site but nonetheless grow as blue or light blue colonies. False negatives most commonly arise from in-frame insertion of DNA fragments containing open reading frames (particularly those of less than several hundred bp). They also result from transcription of the *lacZα* peptide initiated by a promoter within an inserted DNA fragment. Either event results in the synthesis of a foreign peptide fused to the *lacZα* peptide, which may lead to production of blue color. This problem is exacerbated by the high sensitivity of the assay, as very little activity is required to generate blue color (55). These clones are erroneously discarded as nonrecombinants, so the inserted fragments appear difficult to clone. In some cases frequency of false negatives may be very high, constituting up to 25% to 40% of the recombinants (55, and data not shown). Failure to detect these sequences in the blue/white screen may lead to numerous gaps in shotgun DNA sequencing results. The issues of false positives and false negatives apply to direct selection vectors as well. However, false negatives in a direct selection vector are not viable and, thus, cannot be rescued, whereas those in a pUC vector can be recovered by screening blue or light blue colonies for inserts.

Insert-Driven Transcription

Strong promoters can be particularly troublesome to clone in conventional vectors (13, 18). While the molecular mechanisms are not well defined, difficulties in cloning promoters may involve insert-driven transcription interfering with expression of the selectable marker gene or the function of the plasmid's origin of replication (56). Substantial decreases in promoter function and gene expression caused by transcriptional interference are well documented. The presence of two promoters in tandem orientation has been shown to inactivate the downstream promoter in mammalian cells (15, 22), *Drosophila* (11), and bacteria (1, 23). Placing a transcriptional terminator between the two promoters restores activity of the downstream promoter (15, 22). The transcript from a single promoter on a plasmid may even interfere with subsequent reinitiation from the same promoter by reading completely around the plasmid and disrupting the initiation complex. Insertion of a transcriptional terminator into the plasmid prevents the interference (23). Arranging promoters in a convergent orientation results in decreased production of both RNAs. Initiation occurs at full levels, but elongation is strongly inhibited when the transcripts collide (47). These results strongly suggest that cloning promoter-like sequences into pUC plasmids in either orientation may inhibit expression of the vector's drug resistance gene, leading to loss of the clone.

Transcription initiated by cloned promoters is also likely to interfere

Chapter 5. Beyond pUC: Vectors for Cloning Unstable DNA

with plasmid replication. For example, the copy number of plasmids containing an active promoter decreases with increasing activity of the promoter, but only when the transcript is directed toward the origin of replication (34, 57). Insertion of a transcriptional terminator restores the activity of the origin of replication. Read-through from the tetracycline-resistance gene of pBR322 into the origin of replication likewise causes a ninefold decrease in copy number, which can be alleviated by inserting a transcriptional terminator before the origin (56).

High Copy Number

Although the high copy number of pUC plasmids is useful for generating large amounts of plasmid, it decreases the stability of several classes of inserts. Large fragments (e.g., >8 kb) and fragments that are unstable due to previously described sequence or secondary structure can be extremely difficult to maintain in high copy number vectors. Cloning such fragments into low copy number vectors greatly increases their stability (12, 17, 45; and see below).

Ampicillin Selection

The primary disadvantage of ampicillin selection is that the product of the ampicillin resistance gene, beta-lactamase, is secreted by the host cell. As a result, plated colonies create an ampicillin-free zone around themselves, in which nontransformed "satellite" colonies can grow. These satellite colonies interfere with blue/white screening, because they contain no vector and therefore appear to be recombinant white clones. Because of their proximity to true recombinant clones, they also easily contaminate cultures picked from the desired colonies. Similarly, beta-lactamase secreted by recombinants in liquid medium allows growth of cells containing nonrecombinant or deleted plasmids, which often grow considerably faster than the full-length clones. For this reason, a nonsecreted antibiotic, such as kanamycin or chloramphenicol, is often preferred. If an ampicillin resistant vector must be used, the number of satellite colonies on a plate may be reduced by transforming the library into cells of the highest efficiency possible. In this way substantially fewer cells, and hence fewer nontransformed cells, need to be deposited on a given plate to obtain the desired number of transformants, leading to fewer satellite colonies (data not shown).

Plasmid Mobilization

The *nic/bom* region of pBR322 is required in *cis* for the plasmid to be mobilized by other conjugative plasmids (58). Located near the ColE1 origin of replication, parts of the *nic/bom* region are contained in many cloning vectors, including pUC19. To prevent horizontal transfer of pathogenic loci, biosafety concerns or regulations may prohibit cloning of genes from pathogenic organisms into plasmids containing the *nic/bom* site.

Instability of AT-Rich DNA

AT-rich DNA appears to be a uniquely difficult type of substrate to clone. Using pUC plasmids to clone AT-rich genes or genomes [e.g., *B. subtilis* (32), *Plasmodium falciparum* (43), *Dictyostelium discoideum* (21), etc.] often presents extraordinary difficulties. A direct demonstration of the deleterious effect of AT-content is provided with the 5-kb *msp-1* gene of *P. falciparum* (74% AT), which could not be maintained in *E. coli* (43). A synthetic version of the gene, in which the AT content was reduced to 55% and potential bacterial promoter regions were removed, was stably maintained and expressed. While the mechanisms responsible for instability have not been clearly defined, several of the effects described above are likely to increase the difficulty of cloning AT-rich DNA into pUC plasmids.

Numerous AT-rich motifs are associated with bacterial promoters. These elements can occur by chance in cloned AT-rich fragments and may function as promoters in *E. coli*. Transcription from such pseudo-promoters may interfere with plasmid stability, or it may lead to expression of the *lacZα* peptide to generate false negatives (55). For example, randomly selected genomic fragments of *Streptococcus pneumoniae* (overall AT content = 60%) contain elements common to an *E. coli* consensus promoter; notably, these sequences often function as promoters in *E. coli* (39). The presence of such motifs may be responsible for the difficulty of maintaining *S. pneumoniae* clones (39), although this possibility has been disputed (13). A more substantial transcriptional effect may be mediated by motifs similar to the UP element of bacterial promoters. This short segment (AAA a/t a/t T a/t TTTTnnAAAA), or simply a series of A_{5-6} tracts phased at 10-bp intervals (2), is able to stimulate transcription up to several hundred-fold (16), which may destabilize AT-rich clones. The promoter-like activity of AT-rich DNA was directly demonstrated by cloning 1 to 2 kb fragments of *Pneumocystis carinii* genomic DNA into a blue/white screening vector in the absence of IPTG. In this assay, expression of *lacZα* was driven only by the insert DNA. Over 25% of the colonies with inserts produced blue color (data not shown), indicating that reliance on the blue/white screen may obscure a large fraction of true recombinants in a library of AT-rich DNA.

The tendency of AT-rich fragments to mimic plasmid replication origins also may contribute to their instability. Specific AT-rich 13-mers and an adjacent nonspecific AT-rich region are essential for replication in

Chapter 5. Beyond pUC: Vectors for Cloning Unstable DNA   **63**

E. coli (29). A functional hybrid origin can be created from parts of disparate replication origins (54), demonstrating flexibility in recognition of origins. Spontaneous unwinding at the origins (30, 46) and in the neighboring AT-rich region (31, 40) plays a large role in replication. The lack of strict sequence conservation suggests that similar structures or sequences in fragments with high AT content may contribute to plasmid replication. For example, cloning any of numerous fragments from the AT-rich genome of *Clostridium bifermentans* into a low copy number vector resulted in plasmids with unexpectedly high copy number (data not shown). As described earlier, such an increase in copy number *per se* may induce instability. Conversely, spurious origin motifs may in fact induce rearrangement by *inhibiting* replication. In plasmids containing inversely oriented origins, only one origin is active. The other origin is silent, acting as a barrier to the oncoming replication fork, making it susceptible to recombination (50, 51, 63; reviewed in 6).

Vector-driven transcription of AT-rich DNA is another factor that may decrease its stability. A poly(dT)$_{34}$ tract cloned into pUC19 was rapidly lost upon induction of high-level transcription from the *lacZ* promoter (27). The fragment was stable in plasmids containing a mutation that inactivates the *lacZ* promoter, and it was stable in the poly(dA)$_{34}$ orientation, with or without transcription. Spurious deletion of the poly(dT) tract may involve formation of an RNA-DNA triplex, consisting of the nascent poly(U) transcript interacting with the dA-dT duplex. Encountering the triplex causes the transcription complex to stall (28), which leads to instability of the sequence during replication (reviewed in 49). This finding may be of critical importance when attempting to clone fragments containing poly(dA) tracts, particularly when constructing cDNA libraries.

Alternative Vectors

Scores of vectors have been developed to meet the demands of various applications, such as protein expression, protein interaction, mutagenesis, transformation of particular cell types, etc. In contrast, the main vectors for shotgun sequencing—pUC18 and its derivatives—have remained virtually unchanged for over two decades. As most of the aforementioned problems apply to a relatively limited population of DNA fragments, the occasional "unclonable" fragment has been tolerated. However, the impact of these underrepresented sequences becomes significant in whole genome sequencing, complicating the finishing of many genomes and largely preventing assembly of others (32, 45). Recently, several alternative vectors have been developed to reduce the cloning bias inherent to pUC plasmids (Table 5.2).

Table 5.2. Vectors and relevant features for cloning unstable DNAs.

Vector	Size (kb)	Screening Method	Toxic ORFs, AT-rich DNA, Repeats / Vector-driven Expression of Insert Eliminated or Reduced	Short ORFs, Weak Promoters	Promoters, AT-rich DNA / Terminators Flanking Insertion Site	Large inserts, AT-rich DNA / Low Copy Number	Unstable Inserts Lost in Culture / Non-ampicillin Selection	Pathogen DNA / *nic/bom* Site Deleted
pSMART™ LC								
-Kan	2.0	None Req'd.	+	+	+	+	+	+
-Amp	2.0	None Req'd.	+	+	+	+	–	+
pSMART™ HC								
-Kan	1.8	None Req'd.	+	+	+	–	+	+
-Amp	1.8	None Req'd.	+	+	+	–	–	+
pEZSeq™								
-Kan	2.1	Blue/White	–	–[a]	+	–	+	+
-Amp	2.1	Blue/White	–	–[a]	+	–	–	+
pTrueBlue®	2.9	Blue/White	–	+	–	–	–	partial
pTrueBlue®-rop	3.8	Blue/White	–	+	–	+	–	partial
pBlueScript® II	3.0	Blue/White	–	–[a]	–	–	–	–
pGEM®	3.0	Blue/White	–	–[a]	–	–	–	–
pZErO®-2	3.3	Pos. Selection	–	–[b]	–	–	+	partial
pUC18/pUC19	2.7	Blue/White	–	–[a]	–	–	–	partial

[a] Clones containing target insert may be blue (false negative).
[b] Clones containing target insert may be non-viable.

Chapter 5. Beyond pUC: Vectors for Cloning Unstable DNA

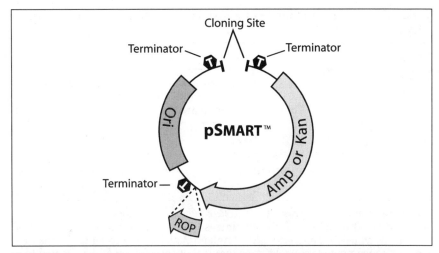

Figure 5.1. **Schematic diagram of pSMART vectors.** The pSMART vectors (Lucigen) contain either ampicillin or kanamycin selection. The *ROP* gene (Repressor of Primer) is present in the low copy versions to decrease copy number. Transcriptional terminators flank the cloning site to prevent insert-driven transcripts from entering the vector. Another terminator follows the drug selection marker to prevent vector-driven transcription of inserts. Vector sizes are 1.8 to 2.0 kb. The pSMART vectors accommodate inserts up to 20 kb.

pSMART

The pSMART series of vectors (Lucigen, Madison, WI) were designed specifically for shotgun library construction (Figure 5.1). They lack several characteristics of pUC18 that have been associated with cloning bias and deletion of inserts. The most notable feature of the pSMART vectors is the absence of transcription into or out of the inserted DNA. Vector driven transcription into the insert is eliminated by removal of the *lacZ* promoter and the *lacZ*α gene. In addition, a transcriptional terminator downstream of the drug resistance gene prevents read-through of insert sequences from this promoter. Because there is no marker gene to indicate which clones have inserts, the pSMART vectors are supplied by the manufacturer pre-cut and dephosphorylated, such that the background of self-ligation is less than 0.1% in most applications. With this low level of background, screening for recombinants is typically not required. The low level of background also facilitates cloning nanogram amounts of sample or cloning samples that produce few colonies for other reasons. Variants of pSMART employ either kanamycin or ampicillin selection, and they contain high or low copy origins of replication.

Extremely efficient cloning in the absence of a visual screen has been verified by sequence analysis of numerous libraries constructed with

Table 5.3. Cloning efficiency of a toxic gene or strong promoter in pSMART or pUC19.

Vector	Target	
	Rnase Gene	**Lambda P_R**
	Total cfu (% Forward)	Total cfu (% Intact)
pSMART-HC	45,000 (38%)	8,500 (75%)
pSMART-LC	20,000 (33%)	3,500 (75%)
pUC19	50,000 (0%)	Blue 100,000 (n.d.)
		White 100 (25%)

Fragments containing a prokaryotic RNase gene or the lambda P_R promoter were amplified with Vent® DNA polymerase (New England Biolabs) and treated with T4 polynucleotide kinase (New England Biolabs), according to the manufacturer's recommendations. Fifty nanograms of each fragment was ligated to blunt, dephosphorylated vector preparations and transformed into electrocompetent *E. coli* cells, using standard molecular biology techniques. Values represent total number of colonies per ligation. PCR was used to determine insert size and percentage of clones in the each orientation.

pSMART. For example, sequencing 32,000 randomly picked clones from a library of 2.5 to 4.5 kb inserts revealed only 40 empty vector clones, or 99.9% recombinants (P. Wilson, Australian Genome Research Facility; personal communication). This level of efficiency is much higher than that obtained with plasmids employing the blue/white screen or *ccd*B selection, which may yield 5% to 10% empty vector clones (i.e., false positives) after screening or selection for recombinants (B. Chiapelli, Washington University Genome Sequencing Center; personal communication).

As expected, the lack of vector-driven expression of insert DNA alleviates the selection against cloning toxic coding sequences. A 350-bp fragment encoding a prokaryotic RNase was recovered in either orientation in pSMART, whereas it could be obtained only in the "reverse" orientation in pUC19 (Table 5.3). Several fragments encoding "toxic" regions of the mouse hepatitis genome also were readily cloned and highly stable in pSMART. These same fragments were very susceptible to rearrangement in standard vectors (70). The lack of transcription is likely responsible for producing more representative cDNA libraries as well (R. Drinkwater, Xenome Corp.; personal communication). Cloning into pSMART vectors is expected to increase the stability of trinucleotide repeats, based on their stability in a transcription-free derivative of pUC (8). Studies are currently underway to verify this prediction.

Transcription initiated by promoters within the insert is blocked by terminators on either side of the cloning site in pSMART (Figure 5.1), thereby insulating essential portions of the vector (i.e., the replication origin and drug resistance gene) from transcriptional interference. The efficacy of this arrangement was demonstrated by cloning a 450-bp fragment containing the strong lambda P_R promoter into pSMART and pUC19. Clones containing this promoter were isolated readily in pSMART, whereas they were rare and nearly obscured by empty vector clones when ligated into a dephosphorylated preparation of pUC19 (Table 5.3).

Recent results indicate that libraries from AT-rich genomic DNA tend to be much larger and more representative when constructed in pSMART rather than pUC19. For example, a cosmid containing genomic DNA from *Pneumocystis carinii* (60%–65% AT) was sheared, end-repaired, and size-selected to 2 to 4 kb. Aliquots of this material were cloned into pSMART-HCKan or pUC19. No deleted inserts or empty vectors were detected among 151 pSMART clones analyzed. In contrast, over 25% of the clones in the pUC19 library suffered obvious deletions, resulting in plasmids smaller than the parental pUC19 (Figure 5.2).

Similar results were obtained with genomic libraries from *Lactobacillus helveticus* (65% AT). Cloning randomly sheared DNA into pSMART-LC produced a large library of intact clones, whereas cloning the same DNA into pUC19 resulted in a small number of clones with a high frequency of deletion (Figure 5.3). Statistical analysis confirmed that the clones from the pSMART library were randomly distributed throughout the genome (Table 5.4).

pEZSeq

The pEZSeq vectors (Lucigen) are closely related to the pSMART series, differing only by the insertion of the *lacZ*α gene to allow blue/white screening for recombinant clones. As with pUC19, transcription into the insert occurs upon induction of the *lac* promoter. pEZSeq is provided by the manufacturer precut and dephosphorylated, with a background of less than 1%. Because of this low background, use of the blue/white screen is not essential. Inactivating the screen (e.g., by using a *lac*Iq strain and no IPTG) may aid in cloning inserts that are unstable or toxic if transcribed.

pSMART VC

The pSMART VC vector (Lucigen) appears to be the most stable vector for cloning extremely recalcitrant DNA fragments, especially those that are over 10 kb in length. This vector incorporates the transcription-free aspects of pSMART into a minimized derivative of pBELOBAC. The pSMART VC vector is single copy for increased cloning stability, but it contains the inducible OriV origin (68) to increase the copy number to 20–100 per cell

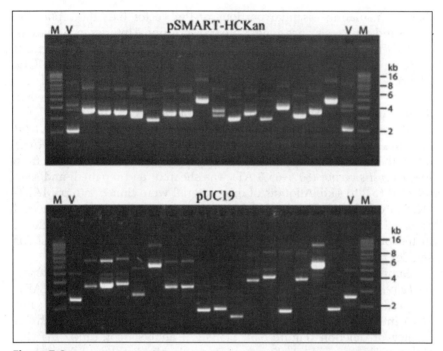

Figure 5.2. Increased stability of AT-rich genomic clones in pSMART. A cosmid containing genomic DNA from *P. carinii* was hydrodynamically sheared, end-repaired (DNATerminator; Lucigen) and shotgun-cloned into pSMART-HCKan or pUC19 (Lucigen). Plasmid DNA was purified from transformants by a standard alkaline lysis protocol. Uncut DNA from each clone was analyzed by gel electrophoresis (approximately 200 ng/lane, 1% agarose, TAE buffer). (upper panel) Plasmids from randomly picked pSMART transformants were all within the expected size range, demonstrating high stability. (lower panel) Over 25% of the pUC19 transformants were unstable, yielding plasmids smaller than the parent vector. M, supercoiled plasmid ladder (Invitrogen); V, empty vector control.

for DNA isolation. This vector facilitated construction of a stable genomic library of 10–20 kb inserts from *Tetrahymena thermophila* (~75% AT), an accomplishment that has not been reported in any other vector (E. Orias, personal communication). Importantly, this DNA was not stable in a similar BAC vector containing the *lacZ*-based blue/white screening.

pTrueBlue

The pTrueBlue vectors (Genomics One, Quebec, Canada) are pUC derivatives designed to eliminate production of false-negative colonies (i.e., blue colonies that have inserts). The multiple cloning site of pUC19 is within a segment of the *lacZα* gene that interferes with expression rather

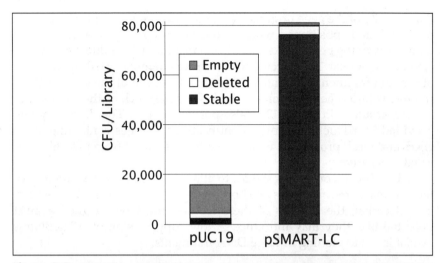

Figure 5.3. Stability of *L. helveticus* genomic DNA in pUC19 and pSMART vectors. *L. helveticus* genomic DNA was hydrodynamically sheared to 2–4 kb, end-repaired (DNA Terminator; Lucigen), and ligated to pUC19 or pSMART-LC (Lucigen). Plasmid DNA from transformants was isolated by alkaline lysis and sequenced with dye terminator chemistry. Sequence analysis of clones from the two libraries showed a 25-fold increase in the number of stable inserts using the low copy number, transcription-free vector pSMART-LC Amp.

Table 5.4. Coverage of *L. helveticus* genome cloned into a pUC-based vector or pSMART.

Library	Clones Sequenced	Bases Sequenced	Genome Equivalents	Sequence Coverage
Sau3A/pJDC9 (pUC-based)	18,400	9 Mb	3.7X	63% (Non-random)
Sheared/ pSMART-LC	7,200	3 Mb	1.4X	69% (Random)
Combined	25,600	12 Mb	5.1X	98%

L. helveticus genomic DNA was partially cut with Sau3A and cloned into pJDC9, using standard techniques. Another sample of the DNA was hydrodynamically sheared (HydroShear, GeneMachines, Inc., San Carlos, CA), end-repaired (DNATerminator, Lucigen), and cloned into pSMART-LC (Lucigen). Statistical analysis indicated that the pSMART library provided the coverage expected from randomly distributed clones.

than activity of the gene product (55). In contrast, the multiple cloning site of pTrueBlue is positioned within a portion of the gene that is essential for activity of the peptide. As a result, cloning into pTrueBlue yields white colonies with near 100% accuracy, which was demonstrated by insertion of random fragments of lambda DNA (55). Insertion of the lambda DNA fragments into other positions of the lacZα gene, such as the site used in pUC19, resulted in up to 42% false-positive clones. The discrimination provided by pTrueBlue allows identification of clones containing small ORFs and weak promoters, which are often discarded from pUC libraries as false negatives.

pTrueBlue incorporates signals to allow high-level expression of proteins from inserts, as well as an f1 origin for generation of ssDNA. As discussed earlier, the benefits of these features must be weighed against potential bias they may introduce. A low copy version of pTrueBlue is available to aid cloning of large DNA fragments.

pZErO

pZErO vectors (Invitrogen) employ direct selection to minimize nonrecombinant clones. They contain a fusion of the lacZα gene to the lethal ccdB gene. Induction of the fused gene results in cell death, unless an insert is present to disrupt the transcript (4). As described earlier, this system greatly reduces the background of empty vector clones. However, short inserts consisting of open reading frames or having promoter activity may allow expression of the ccdB gene, leading to cell death. The pZErO vectors contain the f1 origin of replication, allowing for generation of ssDNA, but this feature also may decrease the stability of certain clones (64).

pTZ, pBlueScript, pGEM

The vectors pTZ (38) (Fermentas, Hanover, MD), pBlueScript (Stratagene), and pGEM (Promega) are closely related to pUC18. The main difference relevant to genomic cloning is the addition of an f1 origin of replication for producing ssDNA. These vectors contain extensive multiple cloning sites, as well as bacteriophage promoters on either side of the cloning site to facilitate making transcripts from the inserts. Their ability to maintain unstable DNA is likely to be nearly identical to that of pUC18.

Conclusions

The vast majority of high throughput sequencing is still carried out with pUC18 and its close relatives. Although these vectors stably maintain most inserts, their limitations are becoming increasingly important with high throughput cloning of targets that may be unstable (e.g., AT-rich, highly repetitive, trace amounts, etc.). Numerous types of inserts are rendered "unclonable" by the high level of transcription and translation caused by the blue/white colony screen. In addition, transcription proceeding through inserts, or initiated within them, may lead to instability of the plasmid or failure of the blue/white screen.

Variants of pUC vectors designed to reduce cloning bias have long been available (62), but no alternative vector has become commonly accepted. Recently developed vectors, designed specifically to circumvent many of the limitations of pUC plasmids, appear to provide more representative libraries for high-throughput cloning and sequencing (Table 5.2). These vectors alter or completely eliminate the blue/white screening system. Vectors that also incorporate transcriptional terminators, lower copy numbers, or resistance to antibiotics other than ampicillin represent the most complete departure from pUC19. Such dedicated cloning vectors substantially reduced the bias against cloning a variety of difficult targets, allowing stable maintenance of strong promoter fragments or toxic coding sequences that were refractory to cloning in pUC-type vectors. The use of low copy number, transcription-free cloning vectors for shotgun cloning resulted in larger libraries with more randomly distributed clones. The effect was most clearly illustrated by cloning AT-rich genomic DNA into the pSMART vectors (Figs. 5.2 and 5.3; Table 5.4). Cloning into transcription-free vectors rather than pUC derivatives appears to substantially decrease the bias in shotgun libraries, facilitating more rapid assembly of genomic sequences.

Acknowledgments

We wish to thank Drs. Jeff Broadbent and Jim Steele for supplying *L. helveticus* genomic DNA and analyzing the resulting libraries (Table 5.4).

References

1. Adhya, S., and Gottesman, M. 1982. Promoter occlusion: transcription through a promoter may inhibit its activity. *Cell* 29: 939–944.
2. Aiyar, S.E., Gourse, R.L., and Ross, W. 1998. Upstream A-tracts increase bacterial promoter activity through interactions with the RNA polymerase alpha subunit. *Proc Natl Acad Sci USA* 95: 14652–14657.
3. Ashley, C.T. Jr., and Warren, S.T. 1995. Trinucleotide repeat expansion and human disease. *Annu Rev Genet* 29: 703–728.
4. Bernard, P., Gabant, P., Bahassi, E.M., and Couturier, M. 1994. Positive-selection vectors using the F plasmid *ccdB* killer gene. *Gene* 148: 71–74.
5. Bichara, M., Pinet, I., Schumacher, S., and Fuchs, R.P. 2000. Mechanisms of dinucleotide repeat instability in *Escherichia coli*. *Genetics* 154: 533–542.

6. Bierne, H., and Michel, B. 1994. When replication forks stop. *Mol Microbiol* 13: 17–23.
7. Bolshoy, A., and Nevo, E. 2000. Ecologic genomics of DNA: upstream bending in prokaryotic promoters. *Genome Res* 10: 1185–1193.
8. Bowater, R.P., Jaworski, A., Larson, J.E., et al. 1997. Transcription increases the deletion frequency of long CTG·CAG triplet repeats from plasmids in *Escherichia coli*. *Nucleic Acids Res* 25: 2861–2868.
9. Bowater, R.P., Rosche, W.A., Jaworski, A., et al. 1996. Relationship between *Escherichia coli* growth and deletions of CTG·CAG triplet repeats in plasmids. *J Mol Biol* 264: 82–96.
10. Cheema, A.K., Choudhury, N.R., and Das, H.K. 1999. A- and T-tract-mediated intrinsic curvature in native DNA between the binding site of the upstream activator NtrC and the nifLA promoter of *Klebsiella pneumoniae* facilitates transcription. *J Bact* 181: 5296–5302.
11. Corbin, V., and Maniatis, T. 1989. Role of transcriptional interference in the *Drosophila* melanogaster Adh promoter switch. *Nature* 337: 279–282.
12. Cunningham, T.P., Montelaro, R.C., and Rushlow, K.E. 1993. Lentivirus envelope sequences and proviral genomes are stabilized in *Escherichia coli* when cloned in low-copy-number plasmid vectors. *Gene* 124: 93–98.
13. Dillard, J.P., and Yother, J. 1991. Analysis of *Streptococcus pneumoniae* sequences cloned into *Escherichia coli*: effect of promoter strength and transcription terminators. *J Bacteriol* 173: 5105–5109.
14. Dabrowski, S., and Kur, J. 1998. Cloning and expression in *Escherichia coli* of the recombinant his-tagged DNA polymerases from *Pyrococcus furiosus* and *Pyrococcus woesei*. *Protein Expr Purif* 14: 131–138.
15. Eggermont, J., and Proudfoot, N.J. 1993. Poly(A) signals and transcriptional pause sites combine to prevent interference between RNA polymerase II promoters. *EMBO J* 12: 2539–2548.
16. Estrem, S.T., Gaal, T., Ross, W., and Gourse, R.L. 1998. Identification of an UP element consensus sequence for bacterial promoters. *Proc Natl Acad Sci USA* 95: 9761–9766.
17. Feng, T., Li, Z., Jiang, W., et al. 2002. Increased efficiency of cloning large DNA fragments using a lower copy number plasmid. *Biotechniques* 32: 992–996.
18. Futterer, J., Gordon, K., Pfeiffer, P., and Hohn, T. 1988. The instability of a recombinant plasmid, caused by a prokaryotic-like promoter within the eukaryotic insert, can be alleviated by expression of antisense RNA. *Gene* 67: 141–145.
19. Reference deleted in proof.
20. Gabrielian, A.E., Landsman, D., and Bolshoy A. 1999–2000. Curved DNA in promoter sequences. *In Silico Biol* 1: 183–196.
21. Glöckner, G., Szafranski, K., Winckler, T., et al. 2001. The complex repeats of *Dictyostelium discoideum*. *Genome Res* 11: 585–594.
22. Greger, I.H., Demarchi, F., Giacca, M., and Proudfoot, N.J. 1998. Transcriptional interference perturbs the binding of Sp1 to the HIV-1 promoter. *Nucleic Acids Res* 26: 1294–1301.

23. Henderson, S.L., Ryan, K., and Sollner-Webb, B. 1989. The promoter-proximal rDNA terminator augments initiation by preventing disruption of the stable transcription complex caused by polymerase read-in. *Genes Dev* 3: 212–223.
24. Iyer, R.R., and Wells, R.D. 1999. Expansion and deletion of triplet repeat sequences in *Escherichia coli* occur on the leading strand of DNA replication. *J Biol Chem* 274: 3865–3877.
25. Jaworski, A., Rosche, W.A., Gellibolian, R., et al. 1995. Mismatch repair in *Escherichia coli* enhances instability in vivo of $(CTG)_n$ triplet repeats from human hereditary diseases. *Proc Natl Acad Sci USA* 92: 11019–11023.
26. Johansen, I.E. 1996. Intron insertion facilitates amplification of cloned virus cDNA in *Escherichia coli* while biological activity is reestablished after transcription in vivo. *Proc Natl Acad Sci USA* 93: 12400–12405.
27. Kiyama, R., and Oishi, M. 1994. Instability of plasmid DNA maintenance caused by transcription of poly(dT)-containing sequences in *Escherichia coli*. *Gene* 150: 57–61.
28. Kiyama, R., and Oishi, M. 1996. In vitro transcription of a poly(dA) × poly(dT)-containing sequence is inhibited by interaction between the template and its transcripts. *Nucleic Acids Res* 24: 4577–4583.
29. Kornberg, A., and Baker, T.A. 1992. *DNA Replication*, 2nd ed. W.H. Freeman. New York.
30. Kowalski, D., and Eddy, M.J. 1989. The DNA unwinding element: a novel, cis-acting component that facilitates opening of the *Escherichia coli* replication origin. *EMBO J* 8: 4335–4344.
31. Kowalski, D., Natale, D.A., and Eddy, M.J. 1988. Stable DNA unwinding, not "breathing," accounts for single-strand-specific nuclease hypersensitivity of specific A+T-rich sequences. *Proc Natl Acad Sci USA* 85: 9464–9468.
32. Krohn, B.M., and Lindsay, J.A. 1993. Cloning of the cyclomaltodextrinase gene from *Bacillus subtilis* high-temperature growth transformant H-17. *Curr Microbiol* 26: 217–222.
33. Kunst, F., Ogasawara, N., Moszer, I., et al. 1997. The complete genome sequence of the gram-positive bacterium *Bacillus subtilis*. *Nature* 390: 249–256.
34. Kwon, Y.S., Kim, J., and Kang, C. 1998. Viability of *E. coli* cells containing phage RNA polymerase and promoter: interference of plasmid replication by transcription. *Genet Anal* 14: 133–139.
35. Lama, J., and Carrasco, L. 1992. Expression of poliovirus nonstructural proteins in *Escherichia coli* cells. *J Biol Chem* 267: 15932–15937.
36. Mardis, E., McPherson, J., Martienssen, R., et al. 2002. What is finished, and why does it matter. *Genome Res* 12: 669–671.
37. Mead, D.A., and Kemper, B. 1988. Chimeric single-stranded DNA phage-plasmid cloning vectors, in Vectors: A Survey of Molecular Cloning Vectors and Their Uses. Rodriguez, R.L., and Denhardt, D.T., eds. Butterworth. Boston, 85–102.
38. Mead, D.A., Szczesna-Skorupa, E., and Kemper, B. 1986. Single-stranded DNA "blue" T7 promoter plasmids: a versatile tandem promoter system for cloning and protein engineering. *Protein Eng* 1: 67–74.
39. Morrison, D.A., and Jaurin, B. 1990. *Streptococcus pneumoniae* possesses canonical *Escherichia coli* (sigma 70) promoters. *Mol Microbiol* 4: 1143–1152.

40. Natale, D.A., Schubert, A.E., Kowalski, D. 1992. DNA helical stability accounts for mutational defects in a yeast replication origin. *Proc Natl Acad Sci USA* 89: 2654–2658.
41. Ohshima, K., Kang, S., Larson, J.E., and Wells, R.D. 1996. Cloning, characterization, and properties of seven triplet repeat DNA sequences. *J Biol Chem* 271: 16773–16783.
42. Ohyama, T. 1996. Bent DNA in the human adenovirus type 2 E1A enhancer is an architectural element for transcription stimulation. *J Biol Chem* 271: 27823–27828.
43. Pan, W., Ravot, E., Tolle, R., et al. 1999. Vaccine candidate MSP-1 from *Plasmodium falciparum*: a redesigned 4917 bp polynucleotide enables synthesis and isolation of full-length protein from *Escherichia coli* and mammalian cells. *Nucleic Acids Res* 27: 1094–1103.
44. Parniewski, P., Bacolla, A., Jaworski, A., and Wells, R.D. 1999. Nucleotide excision repair affects the stability of long transcribed (CTG·CAG) tracts in an orientation-dependent manner in *Escherichia coli*. *Nucleic Acids Res* 27: 616–623.
45. Perng, G.C., Ghiasi, H., Kaiwar, R., et al. 1994. An improved method for cloning portions of the repeat regions of herpes simplex virus type 1. *J Virol Methods* 46: 111–116.
46. Polaczek, P., Kwan, K., and Campbell, J.L. 1998. Unwinding of the *Escherichia coli* origin of replication (oriC) can occur in the absence of initiation proteins but is stabilized by DnaA and histone-like proteins IHF or HU. *Plasmid* 39: 77–83.
47. Prescott, E.M., and Proudfoot, N.J. 2002. Transcriptional collision between convergent genes in budding yeast. *Proc Natl Acad Sci USA* 99: 8796–8801.
48. Quaas, R., McKeown, Y., Stanssens, P., et al. 1988. Expression of the chemically synthesized gene for ribonuclease T1 in *Escherichia coli* using a secretion cloning vector. *Eur J Biochem* 173: 617–622.
49. Rothstein, R., Michel, B., and Gangloff, S. 2000. Replication fork pausing and recombination or "gimme a break". *Genes Dev* 14: 1–10.
50. Santamaria, D., de la Cueva, G., Martínez-Robles, M.L., et al. 1998. DnaB helicase is unable to dissociate RNA-DNA hybrids. Its implication in the polar pausing of replication forks at ColE1 origins. *J Biol Chem* 273: 33386–33396.
51. Santamaria, D., Hernandez, P., Martinez-Robles, M.L., et al. 2000. Premature termination of DNA replication in plasmids carrying two inversely oriented ColE1 origins. *J Mol Biol* 300: 75–82.
52. Schmidt, K.H., Abbott, C.M., and Leach, D.R. 2000. Two opposing effects of mismatch repair on CTG repeat instability in *Escherichia coli*. *Mol Microbiol* 35: 463–471.
53. Reference deleted in proof.
54. Seitz, H., Welzeck, M., and Messer, W. 2001. A hybrid bacterial replication origin. *EMBO Rep* Nov;2(11): 1003–1006.
55. Slilaty, S.N., and Lebel, S. 1998. Accurate insertional inactivation of lacZalpha: construction of pTrueBlue and M13TrueBlue cloning vectors. *Gene* 213: 83–91.

Chapter 5. Beyond pUC: Vectors for Cloning Unstable DNA 75

56. Stueber, D., and Bujard, H. 1982. Transcription from efficient promoters can interfere with plasmid replication and diminish expression of plasmid specified genes. *EMBO J* 1: 1399–1404.
57. Tanaka, M., and Hiraga, S. 1985. Negative control of oriC plasmid replication by transcription of the oriC region. *Mol Gen Genet* 200: 21–26.
58. Tierny, Y., Hounsa, C.G., and Hornez, J.P. 1999. Effects of a recombinant gene product and growth conditions on plasmid stability in pectinolytic *Escherichia coli* cells. *Microbios* 97: 39–53.
59. Twigg, A.J., and Sherratt, D. 1980. Trans-complementable copy-number mutants of plasmid ColE1. *Nature* 283: 2166–2168.
60. Vilette, D., Ehrlich, S.D., and Michel, B. 1995. Transcription-induced deletions in *Escherichia coli* plasmids. *Mol Microbiol* 17: 493–504.
61. Vieira, J., and Messing, J. 1982. The pUC plasmids, an M13mp7-derived system for insertion mutagenesis and sequencing with synthetic universal primers. *Gene* 19: 259–268.
62. Vieira, J., and Messing, J. 1991. New pUC-derived cloning vectors with different selectable markers and DNA replication origins. *Gene* 100: 189–194.
63. Viguera, E., Hernandez, P., Krimer, D.B., et al. 1996. The ColE1 unidirectional origin acts as a polar replication fork pausing site. *J Biol Chem* 271: 22414–22421.
64. Vilette, D., Ehrlich, S.D., and Michel, B. 1996. Transcription-induced deletions in plasmid vectors: M13 DNA replication as a source of instability. *Mol Gen Genet* 252: 398–403.
65. Vilette, D., Uzest, M., Ehrlich, S.D., and Michel, B. 1992. DNA transcription and repressor binding affect deletion formation in *Escherichia coli* plasmids. *EMBO J* 11: 3629–3634.
66. Wang, R.F., and Kuchner, S.R. 1991. Construction of versatile low-copy-number vectors for cloning, sequencing and gene expression in *Escherichia coli*. *Gene* 100: 195–199.
67. Wells, R.D., and Warren, S.T., eds. 1998. Genetic Instabilities and Hereditary Neurological Diseases. Academic Press. San Diego, CA.
68. Wild, J., Hradencna, Z., and Szybalski, W. 2002. Conditionally amplifiable BACs: Switching from single-copy to high-copy vectors and genomic clones. Genome Res 12:1434–1444.
69. Wurtzel, E.T., Movva, N.R., Ross, F.L., and Inouye, M. 1981. Two-step cloning of the *Escherichia coli* regulatory gene ompB, employing phage Mu. *J Mol Appl Genet* 1: 61–69.
70. Yount, B., Denison, M.R., Weiss, S.R., and Baric, R.S. 2002. Systematic assembly of a full length infectious cDNA of mouse hepatitis virus strain A59. *J Virol* 76: 11065–11078.

6 Recombination-Based Cloning

Vincent Ling
Compound Therapeutics, Waltham, MA

Introduction

The power and flexibility intrinsic to recombination-based cloning strategies most likely will allow these new methods to supplant traditional molecular cloning. Whereas traditional molecular cloning methods entail restriction digestion of target DNA followed by ligation into a vector, this procedure necessitates a large collection of restriction enzymes to compensate for incompatibilities that may arise between the cloning vector and desired insert. With recombination-based cloning, the process of molecular cloning has been greatly simplified in that DNA fragments now can be directly "swapped" between different plasmid vectors via recombination reactions, which completely avoids the limitation of incompatible restriction sites, gel purification of target DNA, and ligation reactions. The net result of the gained efficiency of recombination-based cloning methods has lead to the dramatic increase of the accessibility of cloned products, the decrease in time spent on molecular biology projects, and most importantly, the ability to rapidly construct clones in parallel. This chapter outlines the four most popular methods of recombination cloning, with a special focus on the lambda recombinase-based cloning system.

Recombination-Based Cloning: Primary Considerations

There are three components to recombination-based cloning strategies: the donor vector, the acceptor vector, and the DNA fragment usually designated the gene of interest (GOI). The donor vector is usually a plasmid that contains recombinase recognition sites but generally lacks extraneous elements, such as promoters, enhancers, etc., that are used in specific downstream applications. The acceptor vectors contain recombinase recognition sites and necessary DNA elements such as cell-specific promoters, enhancers, polyadenylation sites, etc. Usually, acceptor vectors are generated from traditional expression plasmids by inserting recombination motifs within preexisting multiple cloning sites. Because donor vector and acceptor vectors contain recognition sites, recombination events occur between the two vectors in the presence of the appropriate recombinase. The final form of the recombined product varies depending on the system used but is generally selected by appropriate positive selection.

To enter into the recombination system, the DNA of interest first needs to be inserted into the donor vector either by traditional molecular cloning methods or by a number of alternative nonrestriction enzyme based methods (see below). Once inserted into the donor plasmid, the fragment (if PCR-derived) should be sequenced to ensure the absence of PCR generated errors. Following sequence verification, the donor GOI plasmid is then added with the desired acceptor vectors in a parallel series of recombination reactions, generating expression clones (example given in Figure 6.1).

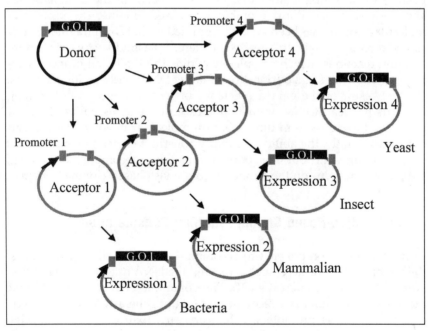

Figure 6.1. Recombination-based cloning allows the mobilization of a gene of interest (GOI) into multiple expression plasmids in parallel. A donor clone containing a GOI is recombined with prepared acceptor plasmids containing desired 5′ and/or 3′ elements necessary for expression or tagging. In this example, the addition of recombinase allows the generation of clones for use in bacteria, mammalian, insect, and yeast expression systems.

Chapter 6. Recombination-Based Cloning

Although a number of recombination systems exist, the most convenient approach, when setting up a moderate throughput or high throughput molecular biology production lab, is to use a commercial system with standardized and optimized reagents. Four types of recombination-based cloning approaches are available and include those based on lambda-phage recombinase (Gateway System; Invitrogen, Carlsbad, CA), the bacteriophage P1 Cre recombinase (Echo System, Invitrogen; and Creator System, Clontech, Palo Alto, CA), and a proprietary bacterial strand-displacement enzyme system (InFusion, Clontech). In terms of conceptual simplicity, the Echo system is perhaps the most direct, with a single-enzyme, single-site approach. The Creator system is slightly more complex, with a single enzyme, double recombination process. The most powerful method is the Gateway system that allows forward and reverse recombination with four sites and two enzyme formulations. Perhaps the most flexible PCR-based method is the InFusion system that allows directional cloning into any site in any vector.

Choosing between a Cre-loxP recombination system and the Gateway recombination system is an important decision that should be made with considerable forethought: These systems are mutually incompatible and are not easily changeable once adopted as a platform cloning system. Each system has unique advantages and disadvantages, and—because of commercial licensing and potential patent reach-throughs—each system carries associated legal ramifications and limitations to their use. Cre recombinase (1) is patented by Steve Ellidge/Baylor University and licensed for distribution by Invitrogen for Echo cloning. Clontech appears to have a separate patent claim for the use of Cre recombinase in the Creator system that may be independent (currently being contested) from Baylor University. At the time of this writing, representatives of neither BD-Clontech nor Baylor University could be reached for comment regarding litigation issues. The Gateway recombination system (2) was initially invented and patented by Life Technologies, but has since been acquired by Invitrogen, and is now marketed under the Invitrogen brand. In early 2004, Invitrogen loosened Gateway licensing restrictions allowing easier distribution of att-site containing clones. Although these products are marketed for research use in both academic and industrial settings, it is suggested that commercial laboratories intending to adopt any recombination-based cloning system do so with full legal evaluation from a corporate counsel.

Another consideration for choosing a recombination system is the availability of preexisting cDNA clones in the desired format. Manufacturers of these systems offer many cDNA libraries and clones preformatted in their own proprietary system. Certain systems also have been adopted by outside institutions that provide clones and libraries. Although a complete list of independent institutional providers in the desired format should be obtained from the manufacturers directly, it is clear that the Gateway system has been adopted by the largest number of consortiums and corporations including: the Japan Government (NEDO-sponsored) human full-length ORF program; the University of Tokyo Schizosacchromyces pombe ORF program; the Dana Farber Cancer Institute *C. elegans* genome and two-hybrid mapping project; the Institut National de la Recherche Agronomique (INRA) Arabidopsis project; the Berkeley Drosophila Genome Project; and MDS Proteomics (FLAG-tagged yeast ORFs). The Harvard Institute of Proteomics has been using the Creator system for DNA clone generation. The NCBI MGC collection contains clones in both Gateway and Creator formats.

In the following sections, the recombination systems are described separately, starting with the recombination strategy first, followed by methods of entry into the recombination system.

Bacteriophage P1 Cre-Based Single loxP Recombination-Echo System

In the Echo recombination system, a single loxP site is present in both the donor and acceptor vectors. The donor vector carries a kanamycin resistance marker whereas the acceptor vector carries an ampicillin resistance marker. In the presence of Cre recombinase, a recombination/fusion event occurs between the two plasmids, and following bacterial transformation, selection of the expression plasmid is achieved by growth on ampicillin plus kanamycin plates. Because this process is based on plasmid fusion, the resultant plasmid is larger than a reaction that transfers only the GOI. For selection purposes, this method also dictates that any acceptor vector generated needs to carry resistance to some antibiotic other than kanamycin, such as ampicillin. Because the fusion product juxtaposes the 5′ promoter of the acceptor vector to the GOI on the donor vector, transcript termination is achieved by the presence of a polyadenylation site directly within the donor vector 3′ proximal to the GOI. A list of currently available acceptor vectors is shown in Table 6.1.

There are two methods to generate Echo-compatible donor clones (Figure 6.2). The first method utilizes traditional molecular cloning approaches to ligate the GOI into an appropriately compatible restriction enzyme sites. The second approach is to PCR amplify the GOI and to react it with prepackaged topoisomerase-linked (TOPO) donor vectors (Uni-1; Invitrogen). TOPO-cloning is a convenient method to subclone GOI independent of internal restriction enzyme sites of either the vector or desired insert. Composition of sequencing primers (provided by manufacturer) 5′ and 3′ to Uni-1 is provided in Table 6.2.

Table 6.1. Echo recombination system acceptor vectors.

Vector	Features
Pichia Expression	
pPICZ-E	AOX1 promoter
pPICZa-E	AOX1 promoter, alpha-factor secretion signal
Insect Expression	
pIB-E	OpIE2 promoter
pBlueBac4.5-E	Polyhedrin promoter (PH)
Mammalian Expression	
pcDNA3.1-E	CMV promoter, Neo selection
pcDNA4/HisMax-E	CMV promoter, SP163 enhancer 5' His6, Xpress, Zeo selection
pcDNA4-E	CMV promoter, Zeo selection
pcDNA6-E	CMV promoter, Bsd selection
pIND-E	Ecdysone inducible promoter
pcDNA4/TO-E	CMV promoter, TetO2 regulation

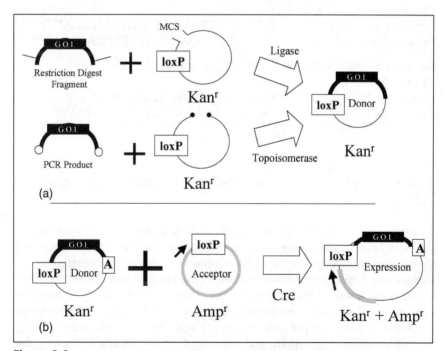

Figure 6.2. Diagram of the single lox-P recombination-Echo system. (a) Insertion of the GOI to generate donor clones may be performed by either traditional restriction digest/ligation cloning methods or by topoisomerase-based fusion into prepared donor vectors. (b) Cre-mediated recombination fusion of donor vector with acceptor vector results in an expression clone. In this system, 3' tags must be incorporated into the GOI-acceptor construct.

Table 6.2. Sequencing primers for Echo, Creator, and Gateway Donor (Entry) vectors.

Echo Donor Vectors (Uni1)
Forward   5'-CTATCAACAG GTTGAACTG-3'
Reverse   5'-CAGTCGAGGC TGATAGCGAG CT-3'

Creator Donor Vectors (pDNR1r; pDNR1, 2, and 3; pDNR-CMV; pDNR-LIB)
Forward (M13)   5'-GTAAAACGAC GGCCAGT-3'
Forward (T7)   5'-ATACGACTCA CTATAGGGC-3'
Reverse (M13)   5'-AAACAGCTAT GACCATG-3'

Creator Donor Vector pDNR-DUAL
Forward (M13)   5'-GTAAAACGAC GGCCAGT-3'
Forward (T7)   5'-ATACGACTCA CTATAGGGC-3'
Reverse (CM1)   5'-AGTACTGCGA TGAGTGGCAG-3'

Gateway Donor and Entry Vectors (pDONR201; pDONR207; pENTR1A, 2B, 3C, 4, and 11)
Forward (near attL1)   5'-TCGCGTTAAC GCTAGCATGG ATCTC-3'
Reverse (near attL2)   5'-GTAACATCAG AGATTTGAG ACAC-3'

Gateway Donor Vectors (pDONR221; pDONRP4-P1R; pDONRP2R-P3)
Forward (M13)   5'-GTAAAACGAC GGCCAG-3'
Reverse (M13)   5'-CAGGAAACAG CTATGAC-3'

Bacteriophage P1 Cre-Based Dual loxP Recombination-Creator System

In a more sophisticated use of Cre recombinase, the Creator system utilizes dual loxP sites in donor vectors that confer ampicillin resistance and sucrose sensitivity (presence of *SacB* gene). During recombination in the presence of Cre, donor clones undergo inter- and intramolecular recombination events with the single loxP acceptor vector resulting in the generation of a dual loxP containing expression clone (Figure 6.3). Bacterial colonies containing expression clones are selected using plates containing ampicillin, chloramphenicol, and sucrose. One interesting feature of the pDNR-dual donor vector is the presence of a splice donor motif immediately downstream of the MCS-GOI insertion site. If an acceptor vector is used that contains a splice acceptor motif downstream of the recombination site, all sequences between the splice signals are edited during transcription, resulting in the loss of the chloramphenicol gene and loxP site

Chapter 6. Recombination-Based Cloning

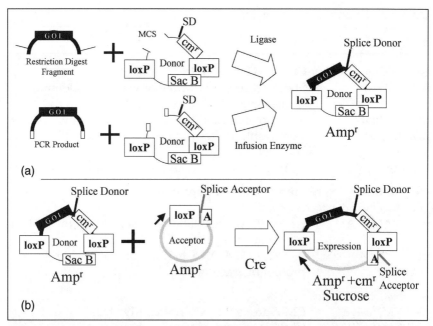

Figure 6.3. Diagram of dual lox-P recombination-Creator system. (a) Insertion of GOI into donor vectors can be accomplished by either traditional cloning or by strand displacement cloning using InFusion enzyme. (b) Recombination of the donor vector with an acceptor vector results in the transfer of the GOI, resulting in the generation of an expression clone. The presence of a splice donor (SD) and a splice acceptor (SA) site in the expression clone allows RNA editing of the transcript, enabling 3' tags (a) to be positioned in the final protein product.

in the final transcript. This splicing event thus allows the fusion of 3' tags embedded within the acceptor vectors. A list of Creator-compatible donor and acceptor vectors are displayed in Table 6.3. Promoter-containing donor expression vectors are available for entry into the Creator system.

To enter the Creator system, two methods are available for generating donor clones. The first utilizes traditional molecular cloning as described above, while the second approach utilizes a novel process of strand displacement using the proprietary InFusion enzyme (Clontech). The InFusion PCR cloning kit provides a linearized donor vector (pDNR-dual) to which the PCR-derived GOI is fused. The only requirement is that the ends of the PCR product have 15 base pairs of sequence complementary to the corresponding proximal 15 base pairs ends on the target vector. The end fusion plasmid undergoes strand displacement reaction and is retrieved upon simple selection of bacterial clones on ampicillin plates. Although the InFusion product is packaged for the Creator system, the InFusion process has the advantage of being widely applicable to many cloning

Table 6.3. Creator system vectors.

Vectors	Features
Generation of Entry Clones (Donor Vectors)	
pDNR-1r	Standard restriction cloning vector for Donor Clone generation.
pDNR-Dual	T7 promoter, conditional splice donor site, his6 tag
pDNR-LIB	Removable stuffer fragment for efficient library construction.
PDNR-CMV	CMV promoter 5' to loxP and multiple cloning site.
Generation of Expression Clones (Acceptor Vectors)	
Yeast two-hybrid expression vectors	
pLP-GADT7 AD	ADH1/GAL4 activation domain fusion for protein interactions
pLP-GBKT7 DNA-BD	ADH1/GAL4 DNA-binding domain fusion for protein interactions
Mammalian expression vectors	
pLP-CMV-Myc	CMV promoter, N-terminal Myc tag
pLP-CMV-HA	CMV promoter, N-terminal HA tag
pLP-CMVneo Acceptor	CMV promoter, neomycin selection
Fluorescent protein fusion vectors	
pLP-EGFP-C1	CMV promoter, N-terminal fusion eGFP (green)
pLP-ECFP-C1	CMV promoter, N-terminal fusion eCFP (cyan)
pLP-EYFP-C1	CMV promoter, N-terminal fusion eYFP (yellow)
pLPS-3'EGFP	CMV promoter, C-terminal fusion eGFP (green)
Bicistronic vectors	
pLP-IRESneo	CMV promoter, internal ribosome entry site, neo selection marker
pLP-IRES2-EGFP	CMV promoter, internal ribosome entry site, eGFP
Tetracycline-regulated expression vectors	
pLP-TRE2	Inducible tet-responsive promoter
pLP-RevTRE	Inducible tet-responsive promoter, retroviral expression vector, hyg selection
Retroviral vectors	
pLP-LNCX	CMV promoter, retroviral expression, neo selection
Adenoviral vectors	
pLP-Adeno-X-CMV	CMV promoter, adenovirus 5 genome

Table 6.3. Creator system vectors. *Continued*

Vectors	Features
Prokaryotic expression vectors	
pLP-PROTet-6xHN	Inducible tet-responsive bacterial promoter, 6xHis tag
Baculovirus expression vectors	
pLP-BakPAK9	Flanking AcMNPV sequences, polyhedron promoter. To be used in subsequent recombination in BakPAK insect cell system
pLP-BakPAK9-6xHN	Flanking AcMNPV sequences, Hisx6 fusion, polyhedron promoter. To be used in subsequent recombination in BakPAK insect cell system.

situations where a properly designed PCR product could be inserted at any linearizable restriction site in any vector (see below). Primers for end sequencing Creator based donor vectors are listed in Table 6.2.

Lambda Phage Recombinase-Gateway System

Without a doubt, the most technically mature recombination system product is the Gateway recombination system, where two enzyme formulations allow exceptional versatility in mobilizing GOI fragments between numerous vectors of choice. Unlike the Cre-based irreversible recombination reaction, the Gateway system allows the user to reversibly mobilize GOI between acceptor and expression clones. This reversible reaction is especially useful when the original donor clone is lost or new modifications are made subsequently in the GOI within the expression clone. Although reversibility appears to be minor advantage at face value, this feature becomes critical in normal laboratory situations (especially in academic settings) where the tracking of clones is oftentimes problematic due to frequent turnover of students, postdocs, and technical staff. In effect, one is not "locked-out" of the recombination system if the original entry clone is lost.

The flexibility associated with Gateway cloning is due to the recombination system used. The primary recombination sites are derived from the lambda phage integration site (att) located on susceptible *E. coli* chromosomes. In the basic Gateway recombination system, sequence derivatives were made of these att sites: attB, attP, attL, and attR (Table 6.4). Two

Table 6.4. Recombination recognition sites.

System	Site	Sequence (5'–3')
Echo, Creator	loxP	GAAGTTAT
Gateway	attB1	ACAAG**TTTGT ACAAAAAAGC** AGGCT
	attB2	ACCAC**TTTGT ACAAGAAAGC** TGGGT
	attP1	AAATAATGAT TTTATTTTGA CTGATAGTGA CCTGTTCGTT GCAACAAATT GATGAGCAAT GCTTTTTAT AATGCCAAC**T TTGTACAAAA AAG**CTGAACG AGAAACGTAA AATGATATAA ATATCAATAT ATTAAATTAG ATTTTGCATA AAAAACAGAC TACATAATAC TGTAAAACAC AACATATCCA GTCACTATGA ATCAACTACT TAGATGGTAT TAGTGACCTG TA
	attP2	AAATAATGAT TTTATTTTGA CTGATAGTGA CCTGTTCGTT GCAACAAATT GATAAGCAAT GCTTTCTTAT AATGCCAAC**T TTGTACAAGA AAG**CTGAACG AGAAACGTAA AATGATATAA ATATCAATAT ATTAAATTAG ATTTTGCATA AAAAACAGAC TACATAATAC TGTAAAACAC AACATATCCA GTCACTATGA ATCAACTACT TAGATGGTAT TAGTGACCTG TA
	attL1	CAAATAATGA TTTTATTTTG ACTGATAGTG ACCTGTTCGT TGCAACAAAT TGATGAGCAA TGCTTTTTTA TAATGCCAAC **TTTGTACAAA AAAGC**AGGCT
	attL2	CAAATAATGA TTTTATTTTG ACTGATAGTG ACCTGTTCGT TGCAACAAAT TGATAAGCAA TGCTTTCTTA TAATGCCAAC **TTTGTACAAG AAAGC**TGGGT
	attR1	ACAAG**TTTGT ACAAAAAAGC** TGAACGAGAA ACGTAAAATG ATATAAATAT CAATATATTA AATTAGATTT TGCATAAAAA ACAGACTACA TAATACTGTA AAACACAACA TATCCAGTCA CTATG
	attR2	ACCAC**TTTGT ACAAGAAAGC** TGAACGAGAA ACGTAAAATG ATATAAATAT CAATATATTA AATTAGATTT TGCATAAAAA ACAGACTACA TAATACTGTA AAACACAACA TATCCAGTCA CTATG

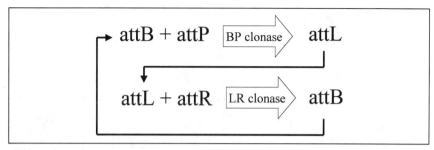

Figure 6.4. Diagram of the basic Gateway recombination reaction system. In the presence of BP Clonase, attB and attP sites recombine to form attL sites. In the presence of LR Clonase, attL and attR sites recombine to reform attB sites.

distinct core 13 nucleotide motifs are present within each of the two sets of matched att sites (attB1, attP1, attL1 and attR1 form one set; attB2, attP2, attL2, and attR2 form the second set). In the presence of the formulated BP Clonase enzyme, matched attB and attP sites recombine to form a matched attL site. Reciprocally, matched attL and attR sites may recombine to reform attB sites in the presence of an alternate enzyme formulation, the LR Clonase mix (Figure 6.4). Thus, depending on the reaction desired, the use of the appropriate att sites and Clonase formulation drives the mobilization of the GOI between donor and acceptor vectors (Figure 6.5).

In a typical LR recombination reaction (Figure 6.6), the GOI is mobilized from the donor plasmid (entry clone) into the acceptor vector (destination vector). The kanamycin-resistant entry clones contain the GOI and are flanked by attL1 and attL2 sites. The ampicillin-resistant destination vector contains a *ccdB* gene (used as a negative selection marker), flanked by attR1 and attR2 sites. Using LR Clonase, recombination between the entry clone (containing attL1, GOI, attL2, Kan^r) and the ampicillin-resistant destination vector (containing attR1, ccdB, attR2, Amp^r) occurs, and the GOI is transferred via an intermediate cointegrant, creating a new expression clone (attB1, GOI, attB2, Amp^r). Presumably, only desired clones with the GOI will be ampicillin resistant, while parental entry clones (Kan^r) would not survive selection and should not propagate due to the negative selection of the toxic *ccdB* gene. This system has been adapted to numerous expression vectors, and ccdB cassette kits are available to convert virtually any vector of interest into a Gateway compatible destination vector. A list of currently available destination vectors is presented in Table 6.5. TOPO-adapted attB expression plasmids are available for direct entry into Gateway acceptor vectors.

Rapid evaluation of Gateway derived clones may be performed by BsrG1 diagnostic restriction digestion, which recognizes all att sites.

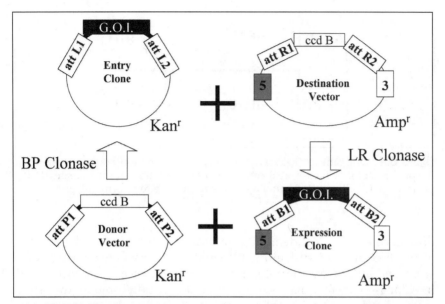

Figure 6.5. Diagram of DNA transfer between plasmids in the Gateway recombination system. On recombination, the GOI in the entry (donor) clone replaces the *ccdB* gene in the destination (acceptor) vector, creating an expression clone. Recombination between an expression clone with a donor vector results in the regeneration of an entry clone.

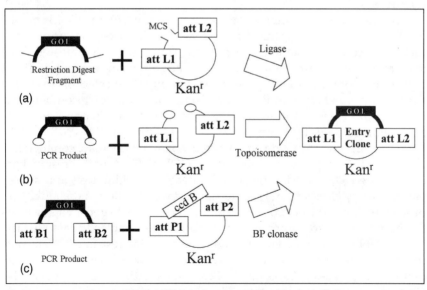

Figure 6.6. Three methods of generating entry clones (attL). (a) Traditional restriction enzyme cloning. (b) Topoisomerase-mediated cloning. (c) BP reaction recombination cloning.

Table 6.5. Gateway vectors.

Vector	Feature
Generation of Gateway Entry Clones (Donor Vectors)	
TOPO TA Cloning	
pCR8/GW/TOPO TA	TA Cloning into Gateway; Spectinomycin selection
Directional TOPO Cloning	
pENTR/D-TOPO	Directional Cloning into Gateway; kanamycin selection
pENTR/SD/D-TOPO	Directional Cloning into Gateway; Shine-Delgarno; kanamycin selection; M13 sequencing primers
pENTR/TEV/D-TOPO	Directional Cloning into Gateway; kanamycin selection; 5' TEV cleavage site
PCR cloning using BP recombination	
pDONR221	Directional Cloning into Gateway; kanamycin selection
pDONRZeo	Directional Cloning into Gateway; zeocin selection
Restriction cloning	
pENTR1A	Vector in reading frame 0; kanamycin selection
pENTR2B	Vector in reading frame +1; kanamycin selection
pENTR3C	Vector in reading frame +2; kanamycin selection
pENTR4	Vector in reading frame 0; modified polylinker from 1A; Kanamycin selection
pENTR11	Kanamycin selection
Generation of Expression Clones (Acceptor Vectors)	
TOPO generation of Expression Clones	
pET160/GW/D-TOPO	T7/*lac* promoter for *E. coli* expression; N-terminal Lumio, 6xHis tags, TEV site
pET161/GW/D-TOPO	T7/*lac* promoter for *E. coli* expression; C-terminal Lumio, 6xHis tags
pcDNA3.2/GW-V5/D-TOPO	CMV promoter for expression in mammalian cells; C-terminal V5; neomycin selection
pcDNA6.2/GW-V5/D-TOPO	CMV promoter for expression in mammalian cells; C-terminal V5; blasticidin selection
pcDNA6.2/cGeneBLAzer-GW/D-TOPO	CMV promoter for expression in mammalian cells; C-terminal *bla*(M) tag; blasticidin selection

Table 6.5. Gateway vectors. *Continued*

Vector	Feature
pcDNA6.2/ nGeneBLAzer™- GW/D-TOPO	CMV promoter for expression in mammalian cells; N-terminal *bla*(M) tag; blasticidin selection

Regulated expression for toxic proteins in E. coli

Vector	Feature
pBAD-DEST49	Inducible expression in *E. coli* from *ara*BAD promoter; N-terminal thioredoxin for improved solubility; C-terminal V6-6xHis for detection and purification.

High-level expression in E. coli

Vector	Feature
pDEST14	T7 promoter; No tag
pDEST15	T7 promoter; N-terminal GST
pDEST17	T7 promoter; N-terminal 6xHis
pDEST24	T7 promoter; C-terminal GST
pET160-DEST	T7/*lac* promoter; N-terminal Lumio, 6xHis tags, TEV site
pET161-DEST	T7/*lac* promoter; C-terminal Lumio, 6xHis tags
pET160/GW/D-TOPO	T7/*lac* promoter; N-terminal Lumio, 6xHis tags, TEV site
pET161/GW/D-TOPO	T7/*lac* promoter; C-terminal Lumio, 6xHis tags
pET104-DEST	T7/*lac* promoter; N-terminal BioEase tag, EK cleavage
pET-DEST42	T7/*lac* promoter; C-terminal V5-6xHis

High-level expression in S. cerevisiae

Vector	Feature
pYES2-DEST52	*GAL*1 promoter; C-terminal V5-6xHis

High-level expression in insect cells

Vector	Feature
BaculoDirect N-term Expression kit	Polyhedrin promoter; N-terminal His-V5 and TEV site
BaculoDirect N-term Transfection kit	Polyhedrin promoter; N-terminal His-V5 and TEV site
BaculoDirect C-term Expression kit	Polyhedrin promoter; C-terminal His-V5
BaculoDirect C-term Transfection kit	Polyhedrin promoter; C-terminal His-V5
pMT-DEST48	Metallothionein promoter; C-terminal V5-6xHis
pDEST8	Polyhedrin promoter; No tag
pDEST10	Polyhedrin promoter; N-terminal 6xHis
pDEST20	Polyhedrin promoter; N-terminal GST

Table 6.5. Gateway vectors. *Continued*

Vector	Feature
Regulated expression in mammalian cells	
pT-REx-DEST30	CMV promoter; No tag
pT-REx-DEST31	CMV promoter; N-terminal 6xHis
Constitutive expression in mammalian cells	
pcDNA-DEST40	CMV promoter; C-terminal V5-6xHis; Geneticin selection
pcDNA-DEST47	CMV promoter; C-terminal GFP; Geneticin selection
pcDNA-DEST53	CMV promoter; C-terminal GFP; Geneticin selection
pcDNA3.1/nV5-DEST	CMV promoter; N-terminal V5; Geneticin selection
pcDNA3.2-DEST	CMV promoter; C-terminal V5; Geneticin selection
pcDNA6.2/V5-DEST	CMV promoter; C-terminal V5; blasticidin selection. Kit includes Tag-on-Demand Technology
pcDNA6.2/GFP/DEST	CMV promoter; C-terminal GFP; blasticidin selection. Kit includes Tag-on-Demand Technology
pcDNA6/BioEase-DEST	CMV promoter; N-terminal BioEase tag and EK site
pcDNA3.2/GW-V5/D-TOPO	CMV promoter; C-terminal V5; neomycin selection
pcDNA6.2/GW-V5/D-TOPO	CMV promoter; C-terminal V5; blasticidin selection
pEF-DEST51	EF-1α promoter for high-level expression from a non-viral promoter; C-terminal V5-6xHis; Geneticin selection
pDEST26	CMV promoter, N-terminal V5-6xHis; Geneticin selection
pDEST27	CMV promoter; N-terminal GST; Geneticin selection
Reporter systems in mammalian cells	
pcDNA6.2/cGeneBLAzer-DEST	CMV promoter; C-terminal *bla*(M) tag; blasticidin selection
pcDNA6.2/nGeneBLAzer-DEST	CMV promoter; N-terminal *bla*(M) tag; blasticidin selection
pcDNA6.2/cGeneBLAzer-GW/D-TOPO	CMV promoter; C-terminal *bla*(M) tag; blasticidin selection

Table 6.5. Gateway vectors. *Continued*

Vector	Feature
pcDNA6.2/ nGeneBLAzer-GW/ D-TOPO	CMV promoter; N-terminal *bla*(M) tag; blasticidin selection
pcDNA6.2/ cLumio-DEST	CMV promoter; C-terminal Lumio tag; blasticidin
pcDNA6.2/ nLumio-DEST	CMV promoter; N-terminal Lumio tag; blasticidin
Lentiviral Expression Vectors	
pLenti6/V5-DEST	CMV promoter; C-terminal V5; blasticidin selection
pLenti4/V5-DEST	CMV promoter; C-terminal V5; Zeocin selection
pLenti6/UbC/ V5-DEST	UbC promoter; C-terminal V5; blasticidin selection
pLenti4/TO/V5-DEST	CMV/TO promoter; C-terminal V5; Zeocin selection
pLenti6/TR	CMV promoter; β-globin IVS-*tet*R; blasticidin selection
pLenti6/ BLOCK-iT-DEST	Requires special entry clone, with U6 pol III promoter; C-terminal V5; blasticidin selection
BLOCK-iT U6 RNAi Entry Vector	U6 promoter; pol III terminator
Adenoviral Expression Vectors	
pAD/CMV/V5-DEST	CMV promoter; C-terminal V5; TK polyA
pAD/ BLOCK-iT™/ V5-DEST	Requires special entry clone, with U6 pol III promoter
Gene Targeting vectors	
pEF5/FRT/V5-DEST	EF-1α promoter; C-terminal V5
In Vitro *Protein Synthesis*	
pEXP1-DEST	T7 promoter; Shine Dalgarno; N-terminal 6xHis; EK cleavage
pEXP2-DEST	T7 promoter; C-terminal V5-6xHis
pEXP3-DEST	T7 promoter; N-terminal Lumio tag and 6xHis
MultiSite Gateway Three-Fragment Vector Construction Kit	
	pDONR P4-P1R
	pDONR P2R-P3
Library construction	
	pCMVSPORT6 Not I/Sal I digested
	pEXP-AD502 Not I/Sal I digested

Chapter 6. Recombination-Based Cloning

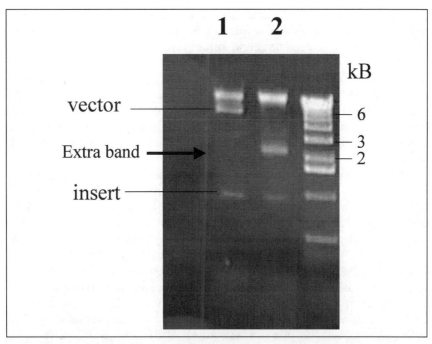

Figure 6.7. Example of plasmids derived from LR reactions. Following BsrG1 diagnostic digestion, plasmid samples were fractionated by 1% agarose gel electrophoresis. Successful LR reaction releases the expected insert band (Lane 1). Unsuccessful LR reaction yields an expected an additional ~2.2 kb band (Lane 2). Additional analysis reveals bacterial clones from unsuccessful LR reaction to be both ampicillin and kanamycin resistant.

Digestion of DNA minipreps of either donor or acceptor clones excises the GOI fragment from the Gateway vector backbone. We find that the LR reaction to be generally robust for most reactions, typically being over 85% efficient, with the remaining 15% generating some yet-to-be-determined intermediate cloning artifact. This artifactual plasmid by-product produces a 2.2 kb fragment upon BsrG1 digestion (Figure 6.7) that may or may not be in molar equivalent with other vector and fragment bands within that sample. Usually, finding a successful recombinant clone is achieved after the analysis of three colonies. We find that clones containing the 2.2 kb fragment to be correlated with dual resistance to ampicillin and kanamycin. We also have noted that the number of cointegrants increase with respect to the continued handling of the thermolabile LR Clonase above −80°C, and with the degree of repetitive sequence composition within the destination vector. Destination vectors with long stretches of repetitive sequences such as retroviral and lentiviral vectors are more susceptible to cointegrate formation, even when fresh LR

Clonase is used. In these incidences, a single/double antibiotic colony screening strategy should be implemented to determine whether colonies picked are ampicillin resistant and kanamycin sensitive. Indeed, we have found the first release of the Invitrogen lentivirus destination vector kit (pLenti6V5-DEST) to produce nearly 80% false positives upon LR reaction, with the true positive clones (Amp^r, Kan^s) corresponding to the smaller colonies on the selection plates. Technical updates are available from Invitrogen regarding the use of an alternate HB101 *E. coli* strain derivative, Stbl3, to reduce the number of false positives generated with repeat-rich, susceptible destination vectors.

There are currently three ways to generate entry (donor) clones in the Gateway system. As with the other systems previously described, traditional molecular cloning methods may be employed to ligate a GOI into the multiple cloning site (MCS) of Gateway entry vectors (pENTR vector series). An alternate method would be to use directional topoisomerase based cloning, where the GOI is transferred into the pENTR-TOPO series of vectors, either in single reaction format or in plate-based high throughput format. Both series of pENTR vectors confer kanamycin resistance and also contain attL1 and attL2 sites. The third method is to recombine the GOI into a donor vector using BP Clonase. The GOI is prepared by PCR with oligonucleotide primers encoding attB1 and attB2 sites. The attB1/attB2 adapted GOI is then recombined with a donor vector (conferring kanamycin resistance) encoding a *ccdB* gene flanked by attP1 and attP2 sites. BP recombination reactions with a mixture of the fragment and donor vector results in the generation of entry clones containing the GOI flanked by attL1 and attL2 sites. BP reactions are extremely robust and produce clones virtually free from artifactual recombination intermediates. To insure the sequence consistency of the PCR product, DNA sequencing should be performed on PCR-derived entry clones. For reasons undetermined, earlier pDONR201 based entry clones were plagued by sequencing difficulties stemming from the lack of good vector-primer binding sites that would allow efficient and lengthy end-reads of the GOI. To bypass this problem, pDONR221 donor vector was introduced containing standard M13 forward and reverse primer sites from which adequate sequence reads may be reproducibly generated. Primers useful for Gateway vector sequencing are listed in Table 6.2.

The Future of Recombination Cloning

The Gateway system is clearly the early leader in the field of recombination cloning with its substantial user-base and continued support and innovation in the Gateway product line. Indeed, new att sites derivatives (attB3, attP3, attL3, and attR3; attB4, attP4, attL4, and attR4) have been introduced, allowing even greater flexibility in recombination cloning (Multi-site Gateway; Figure 6.8). Multi-site Gateway allows the generation of custom plasmids with 5' and 3' elements. The process starts with the generation or collection of the first set of entry clones contain 5' elements (promoters) flanked by attL4 and attR1 sites. The second set of entry clones, supplied by the user, would contain GOI (ORF) flanked by attL1 and attL2. The third set of entry clones would contain different 3' elements (tags and poly-A sites) flanked by attR2 and attL3. When four-way recombination using modified LR-plus recombinase is performed, the addition of the three entry clone to the acceptor vector pDEST-R4-R3 generates a resultant expression plasmid containing [attB4]-5' element-[attB1]-GOI-[attB2]-3' element-[attB3]. With sufficiently large libraries of different 5' elements and 3' elements, expression clones of any promoter and 3' tag combination theoretically can be generated, potentially ending the need for specific single-purpose acceptor vectors. Despite the power of Multi-site Gateway to produce customized plasmid configurations, it is yet to be determined whether this product will be widely adopted by current Gateway users because of the additional complexity associated with this very flexible system.

The InFusion strand displacement method (Figure 6.9), currently used as an entry system in the Creator system, offers an alluring alternative to recombination cloning by its potential to easily insert any fragment to any vector. Unlike restriction or recombination processes where recognition sites are needed, the sole requirement of the InFusion system is that the 15 base pair ends between fragment and vector are complementary. Currently, the components of the InFusion extract are proprietary and its mechanism of action has not been disclosed, so it is also not currently known whether the current InFusion formulation is sufficiently robust to be widely used for purposes other than entry into the Creator vector system. However, in our hands, we have been able to PCR-clone attB1-GOI-attB2 fragments into inversely amplified Gateway expression vectors (pcDNA-DEST40 and pEF-DEST51) simply based on attB complementary primers. Screening of 96-well plate–based clones for gene function revealed 96% efficiency of cloning (unpublished data). These results strongly suggest that the InFusion system can be used to clone any user-defined PCR product into any corresponding inversely amplified expression vector, thereby removing any need for predefined recombination or restriction sites that include unwanted extraneous sequences. Like all PCR-fragment assemblies, care must be taken to sequence and verify appropriate segments of both the parental vector and GOI to assure that mutations are not propagated through this process. However, if used as an activity screen of large cDNA or domain libraries, the InFusion strand system offers clear advantages over the other previously described recombination systems, especially in its currently packaged lyophilized room

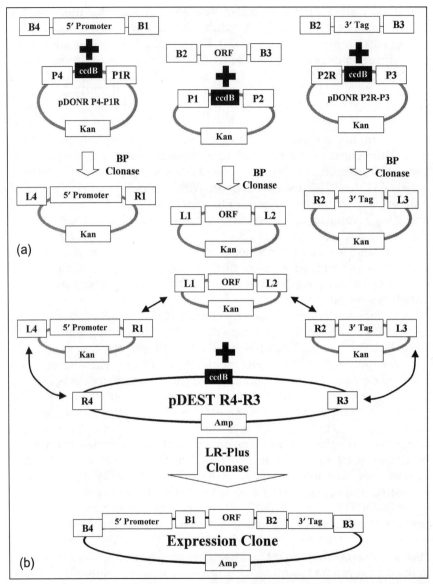

Figure 6.8. Multi-site Gateway: A component based plasmid recombination system. (a) Component 5′ element, 3′ element, and GOI are sub-cloned or mobilized into modified entry vectors. (b) Recombination between the three plasmids and pDEST-R4-R3 results in the combinatorial generation of the desired expression vector.

Chapter 6. Recombination-Based Cloning

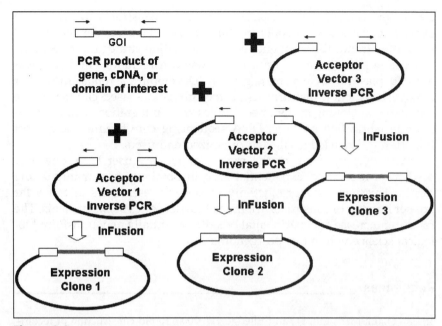

Figure 6.9. InFusion generation of plasmid expression clones. A gene, cDNA, or domain of interest is amplified by PCR with user-defined 5′ and 3′ oligonucleotide overhangs. Acceptor vectors are prepared by inverse PCR with >15 bp complementary 5′ and 3′ nucleotide overhangs. InFusion enzyme action causes strand displacement resulting in directional cloning of GOI, resulting in formation of expression clones.

temperature form. Because the InFusion system is an irreversible cloning method, subsequent transfer of GOI between plasmids is not possible unless integrated with plasmids already in either the Gateway or Creator recombination system.

It is fairly reasonable to presume that within the next few years, a leading recombination "operating system" will emerge between the competing products. The dominant molecular cloning system will likely become a premium product that leads the industry due to the synergy between increasing use and expanding user-base. Prominence in this field will be bolstered by clear and accurate documentation, reagent quality control, and reasonable marketing/licensing of proprietary enzymes and processes. Within the communications and technology industries, the power of connectivity products is directly related to the size of the user-base. The increased convenience of information transfer between disparate end-users exponentially enhances utility. Likewise, the emergence of a dominant recombination cloning system will transform the art and technique of molecular biologist from the model of "cottage-industry craftsman" sculpting low numbers of cDNA constructs by restriction cloning to the modern model of standardized data transfer using PCR assembly, strand displacement and recombination strategies. Indeed, the increase in plasmid throughput raises new considerations in data tracking and plasmid archive management. Developing a robust molecular biology production platform using recombination-based cloning as the foundational cloning system necessitates close integration with oligonucleotide fabrication services, DNA sequencing core facilities, and most importantly, knowledgeable and organized end-users.

With the expansion of the molecular biology user base, one may predict an associated exponential ease in obtaining and manipulating clones. This increased productivity empowers researchers to test a far greater number of constructs in a much broader series of experiments. The resulting increase of experimental breadth potentially may transform biological science within this postgenomic era.

References

1. Liu, Q., Li, M.Z., Liu, D., and Elledge, S.J. 2000. Rapid construction of recombinant DNA by the univector plasmid-fusion system. *Methods Enzymol* 328: 530–549.
2. Walhout, A.J., Temple, G.F., Brasch, M.A., et al. 2000. GATEWAY recombinational cloning: application to the cloning of large numbers of open reading frames or ORFeomes. *Methods Enzymol* 328: 575–592.

7 Plasmid Preparation Methods for DNA Sequencing

Parke K. Flick
454 Life Sciences, Branford, CT

Introduction

Successful sequencing of plasmid-derived DNA depends strongly on the purity and structural integrity of the DNA. For this reason, considerable effort has been expended over the past 15 years or so to develop methods that are both convenient and efficacious for the extraction and purification of plasmid DNA. As DNA sequencing has become more automated and high-throughput, these methods have evolved to keep pace. Beginning with low-throughput methods involving numerous manipulations and centrifugation steps, the technology has advanced to the stage where liquid handling robots are able to carry out pipetting steps in conjunction with magnetic beads that bind the DNA and obviate the requirement for centrifugation or vacuum elution. Purifications may now be carried out in 96- or 384-well plates for high-throughput applications. Most of these methods involve the use of efficient binding matrices for separating DNA from proteins and other contaminants. The matrix can be contained within the wells of the microtiter plate or packed in spin columns for processing smaller numbers of samples.

The purpose of this chapter is to review the best current methods for preparing plasmid DNA for sequencing. Emphasis will be placed on a new amplification method for circular templates, TempliPhi, which eliminates completely the need for culturing and subsequent purification steps.

Methods for Low Throughput Preparation of Plasmid DNA

Virtually all plasmid preparation methods employ a modification of the alkaline lysis procedure (2, 3). Bacterial cells from an overnight culture of

1 to 3 ml are pelleted and resuspended in a lysis buffer which lyses the cells. A neutralization solution is then typically added and a cleared lysate prepared by centrifugation or filtration. The lysate, containing the plasmid DNA of interest, is then processed in a variety of ways depending on the commercial supplier. Commonly, it is passed through a silica-based matrix in a spin column or multi-well plate under conditions where the binding of DNA to the matrix is favored (high salt, chaotropic agent). Following a washing step with medium-salt buffer to get rid of proteins and other bound material, the plasmid DNA is eluted from the matrix with low-salt buffer or distilled water. In many cases the DNA is sufficiently concentrated to use as is, although a final alcohol precipitation step may be needed to raise the concentration to desired levels.

There are many variations on the above theme. For example, a number of suppliers employ a silica matrix in their products (Bio-Rad, Richmond, CA; Rapid Concert, Invitrogen, Carlsbad, CA; Eppendorf, Westbury, NY; Stratagene, La Jolla, CA; Promega, Madison, WI; and QIApreps, Qiagen, Valencia, CA), whereas others use a filter matrix (Whatman, Clifton, NJ; Invitrogen, Concert 96/384; Edge Biosystems, Gaithersburg, MD; and Millipore, Billerica, MA). Fluids are passed through the matrix either by the use of spin columns (Bio-Rad, Invitrogen Rapid Concert, Stratagene, Qiagen QIApreps) or vacuum (Promega Wizard Plus, Qiagen QIAprep 96 Turbo, Whatman, Millipore), although Qiagen and Invitrogen also offer products that use gravity-flow anion exchange columns for the preparation of high-purity DNA.

Methods for High Throughput Preparation of Plasmid DNA

High throughput plasmid DNA preparation (96 samples or greater) takes place in microtiter plates. Individual colonies are picked and inoculated into the wells of a growth plate or flat-bottom block each containing approximately 1.0 ml of media plus antibiotic and grown up overnight (see also Chapter 8 for additional information). Bacteria are then typically pelleted by centrifugation for 10 minutes, the supernatant removed, and the pellets resuspended in buffer. An aliquot of each resuspended pellet (100–200 µl) is then transferred to a second, filter plate containing the binding matrix. Lysis solution is added and the mixture either centrifuged through the matrix into the wells of a receiver plate containing alcohol to precipitate the plasmid DNA or transferred by means of vacuum through the matrix into the receiver plate. Following another centrifugation step to collect the DNA, the pellets are resuspended in distilled water or dilute buffer. Again, there are numerous variations on this theme. A key one is the use of magnetic beads to bind DNA, allowing contaminants to be separated by the application of a magnetic field, followed by washing. Purified plasmid DNA is then eluted from the beads with water. No centrifugation or vacuum filtration is required. Agencourt Bioscience Corporation (Beverly, MA) has introduced SPRI technology (solid phase reversible immobilization) for this purpose, which is based on published work by Hawkins et al. (10). In their MCPrep 96-well plasmid purification system, SPRI beads, which are paramagnetic beads modified with carboxylate groups on a surface coating, are mixed with resuspended bacterial pellets and a special lysis buffer in a 96-well plate. The solution lyses the cells and causes bacterial chromosomal DNA and cellular debris to selectively bind to the beads. A magnetic plate is then used to bring the beads to the bottom of the well. The plasmid-rich supernatant is then transferred to the wells of another plate containing SPRI beads in a second solution that causes plasmid DNA to bind to the beads. After a suitable incubation time, a magnetic plate causes the beads to collect near the bottom of the wells. The supernatant is discarded, the beads dried briefly and washed with 70% ethanol, and the plasmid DNA eluted with water or low-salt buffer. Then it can be transferred to another plate for use. This methodology is very amenable to automation using robotic workstations, and it has been successfully applied to BAC end sequencing and sequencing reaction clean-up in addition to plasmid preparation (12, 14).

Table 7.1 summarizes the product offerings for plasmid DNA preparation of a number of representative companies. Included is a brief description of the technology along with an estimate of the final yield of plasmid DNA from 1 ml of culture.

TempliPhi: A New Template Preparation Method without Culturing and Purification

In 2001, Amersham Biosciences (Piscataway, NJ) introduced TempliPhi, a novel isothermal amplification method for circular DNA templates. Using this method, it is possible to start with as little as 100 bacterial cells or 1 µl of an overnight liquid culture for amplification. Glycerol stocks up to 1 µl also can be used. Amplification is performed using a unique DNA polymerase from *Bacillus subtilis* bacteriophage phi29 (6). This single-subunit enzyme possesses an extremely high processivity that is capable of incorporating at least 70,000 nucleotides in one binding event (4, 13). In addition, the enzyme has strand displacement activity and an associated 3'–5' proofreading exonuclease activity (5, 9), which allows for high fidelity replication of input DNA templates. The error rate of the polymerase is estimated to be 3×10^{-6} (9).

The mechanism of the TempliPhi amplification reaction is rolling circle amplification (RCA) by extension and strand displacement of random sequence hexamer primers (7). This reaction is illustrated in

Table 7.1. A representative list of products for preparation of plasmid DNA for sequencing.

Supplier	Products	Technology	Yield of DNA[a]
Agencourt Bioscience (www.agencourt.com)	MCPrep™ 96 Well Plasmid Purification	Alk. Lysis-coated paramagnetic beads (SPRI™) for clearing and binding plasmid DNA. Elution with water.	1–2 µg/ml culture (for high copy-number plasmids only)
Bio-Rad Laboratories (www.bio-rad.com)	Quantum Prep®	Silica matrix-spin columns	Up to 25 µg/ 1.5 ml culture (rich broth)
Edge BioSystems (www.edgebio.com)	Plasmid 96 Miniprep Kit	Alkaline lysis-96-well filter plate-centrifuge into alcohol in receiver plate	5 µg/ml culture
Eppendorf/ Brinkmann (www.eppendorf.com)	Perfectprep® mini- 1–3 ml midi- 10–75 ml maxi- 100–300 ml	Enhanced alkaline lysis-binding and elution from glass matrix	5–10 µg/ml culture
Invitrogen/Life Technologies (www.invitrogen.com)	Concert™ Rapid Plasmid System—mini-, midi-, and maxiprep	Spin cartridge-silica-based membranes	5 µg/ml culture
	Concert™ High Purity Plasmid Purification System	Anion exchange resin column	5 µg/ml culture
	Concert™ 96/384 Plasmid Purification Systems	Solid phase lysis-filter matrix in wells of plate-plate centrifugation	2–5 µg/well = 10–25 µg/ml culture
Millipore (www.millipore.com)	Montage™ Plasmid Miniprep96 Kit	96-well–three filtration steps using vacuum. Plasmid DNA retained on membrane filter	4–6 µg/ml culture
Promega (www.promega.com)	Wizard® Plus Minipreps DNA Purification System	1–3 ml culture-alkaline lysis-binding resin in minicolumn-vacuum or spin elute	3–5 µg/ml culture

Table 7.1. *Continued*

Supplier	Products	Technology	Yield of DNA[a]
	Wizard® SV96 Plasmid DNA Purification System	96-well–clearing plate + binding plate. Vacuum elution	3–5 µg/ml culture
	Wizard® Magnesil™ Plasmid Purification System	Multiwell-paramagnetic particles for clearing and plasmid capture	5–8 µg/ml culture
Qiagen (www.qiagen.com)	Qiagen Spin Columns	Individual columns—silica matrix	Up to 20 µg from 1–5 ml culture
	QIAprep™ Spin Miniprep	Modified alk. Lysis-spin column-bind DNA to membrane with chaotropic salt	Up to 20 µg from 1–5 ml culture
	QIAprep™ 96 Turbo Miniprep (high throughput)	Modified alk. Lysis-filter plate, then capture plate with membrane. Elute by vacuum.	Up to 20 µg from 1–5 ml culture
	QIAwell™ 96 Ultra Plasmid Purification Kit (high purity, high throughput)	Additional plate absorption step (QIAwell plate) between filter plate and prep plate	4–5 µg/ml culture
Stratagene (www.stratagene.com)	StrataPrep®	Modified alkaline lysis-microspin cup with silica-based fiber matrix	5–20 µg/ml culture (cup binding capacity is 20 µg)
Whatman (www.whatman.com)	96-well Plasmid Miniprep Kit	Alkaline lysis-clarification plate-DNA binding plate	5 µg/ml culture
	High throughput plasmid binding plate	96 × 800 µl wells-filter binds DNA-vacuum elution	6 µg/well max.

[a] Yield is based on high copy-number plasmids.

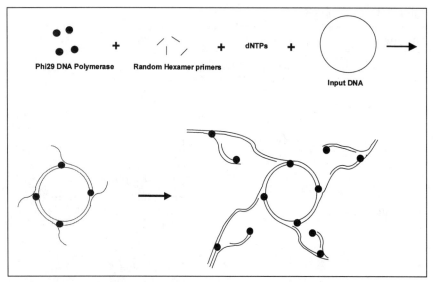

Figure 7.1. Schematic of the TempliPhi process. Random hexamer primers anneal to the circular template DNA at multiple sites. Phi29 DNA polymerase extends each of these primers. When the enzyme reaches the 5' end of a downstream primer, strand displacement occurs and the enzyme continues to copy around the circle. Displaced strands are available for additional priming by more hexamers. The process continues, resulting in a large, highly branched structure and giving exponential amplification.

Figure 7.1. The random sequence hexamers contain phosphorothioate modifications at their 3' ends, rendering them resistant to the 3'–5' exonuclease activity of phi29 DNA polymerase and allowing continued priming events to occur during extended DNA amplification reactions (7, 11). The input DNA can be a plasmid or any other circular piece of DNA, such as bacteriophage M13 clones, cosmids, fosmids, or BAC DNA. Interestingly, linear DNA, such as human chromosomal DNA, also can be amplified by this same general method, although there is some minor loss of representation from the ends in the final product (8, 15). Further discussion of linear DNA amplification is beyond the scope of this chapter. A product for this specific purpose will be introduced by Amersham Biosciences in 2003.

The DNA can be added to the TempliPhi reaction as such or in the form of bacterial cells or phage plaques containing the clone of interest. Reactions are carried out at 30°C for 4 to 12 hours. Longer reaction times may be used without compromising yield or quality of amplified DNA. The product of the TempliPhi reaction is high molecular weight, double-stranded concatemers of the starting circular template (Figure 7.1). Ampli-

Chapter 7. Plasmid Preparation Methods

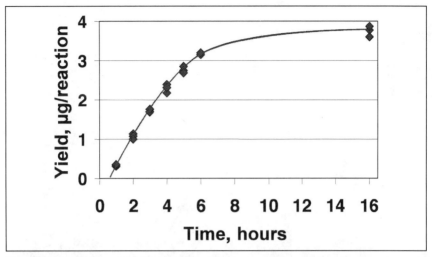

Figure 7.2. **Typical kinetics of the TempliPhi reaction.** One nanogram of pUC18 DNA was heated to 95°C for 3 minutes in 10 µl of TempliPhi Denaturation Buffer. The solution was then chilled on ice, 10 µl of TempliPhi Premix added, and the reaction incubated at 30°C for 16 hours. At the indicated times, amplified DNA was quantified (in triplicate) using PicoGreen reagent (Molecular Probes, Eugene, OR).

fication factors of up to 10^7 are commonly obtained, meaning that microgram quantities of product can be obtained from picogram quantities of input DNA. Typical kinetics for the TempliPhi reaction is shown in Figure 7.2. The key advantage of TempliPhi technology is that it eliminates the need for overnight culture growth of bacteria/phage, saving at least one day of preparation time before sequencing. As an added convenience, the amplified DNA can be used directly in sequencing reactions or other applications without precipitation or further clean-up.

Figure 7.3 presents a brief protocol for Templiphi. A small sample (1 µl or less) of material to be amplified is added to 5 µl of the Sample Buffer. The mixture is heated to 95°C for three minutes and then cooled to room temperature. Five microliters of a mixture of Reaction Buffer and Enzyme Mix are added to the cooled mixture, and the complete reaction is incubated at 30°C for 4 to 12 hours. The enzyme is then inactivated by heating to 65°C for 10 minutes. This amplified DNA may be used immediately or stored at −20°C.

Figure 7.4 shows a characterization of the products of TempliPhi reactions as a function of time, amount of input DNA, and source of the DNA. The template in this case is the plasmid pUC18. Note that for four hours of amplification, as little as 15 pg of input purified pUC18 yields

Short Protocol:
1. Add a small sample (1μl or less) of material to be amplified to 5μl Sample Buffer (white cap). Mix well.
2. Cap container, heat to 95°C for 3 minutes, and cool to room temperature.
3. Add 5μl of combined Reaction Buffer (blue cap) and Enzyme Mix (yellow cap) (5μl Reaction Buffer + 0.2μl Enzyme Mix per sample).
4. Incubate at 30°C for 4 to 12 hours.
5. Inactivate enzyme by heating to 65°C for 10 minutes (optional).
6. Store amplified material at −20°C.

TempliPhi Workflow

Liquid culture/M13 Phage culture supernatant
Glycerol stock

Bacterial colony/M13 plaque

Step 1

Transfer a small amount (1μl or less) of liquid culture or small portion of a colony or ~1 ng of purified DNA to 5μl sample buffer

5μl Sample Buffer + input DNA

Step 2 Heat 95°C for 3 minutes, cool to room temperature or chill on ice.

Step 3 Mix 50 μl of Reaction Buffer and 2 μl Enzyme Mix in a separate tube on ice. (sufficient for 10 TempliPhi reactions)

Step 4 Transfer 5 μl of the above Reaction Buffer + Enzyme Mix to each tube from step 2. Mix gently. Perform TempliPhi amplification reaction at 30°C for at least 4 hours.

Step 5 Perform heat inactivation at 65°C for 10 minutes (optional). Transfer 1μl of the product directly into a standard cycle sequencing reaction.
Perform cycle sequencing with suitable sequencing primers, post sequencing clean up and analyze sequencing products in appropriate electrophoresis system.

Figure 7.3. Brief protocol for the use of TempliPhi.

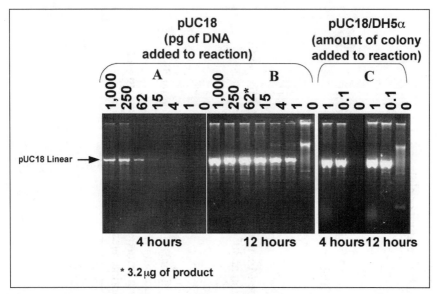

Figure 7.4. Characterization of the products of the TempliPhi reaction and yield as a function of amount of input DNA. Either purified pUC18 DNA (A and B) or pUC18 DNA contained in colonies of *E. coli* DH5α (C) was used as the starting template. TempliPhi reactions were carried out with the indicated amount of input template for the time indicated at the bottom of each panel. Following incubation, each reaction mixture was heated to 65°C for 10 minutes to inactivate the enzyme and the amplified DNA was ethanol-precipitated, resuspended in *Eco*RI digestion buffer, and digested with *Eco*RI for one hour at 37°C. One microliter of the *Eco*RI digest was then loaded onto a 1.0% agarose gel, which was run and stained with ethidium bromide. The right-most lane in each panel is the control with no input template. The total amount of amplified product obtained from 62 pg of input template after 12 hours (B) was measured and found to correspond to 3.2 μg of product. Minor DNA bands visible on the gel most likely arise from non-specific amplification of hexamers present in the reaction mixture (T. Mamone, personal communication).

detectable product as determined by agarose gel electrophoresis of a *Eco*RI restriction digest of the reaction (Figure 7.4A). With 12 hours of amplification, 1 pg of input pUC18 yields a significant amount of product (Figure 7.4B). It should be noted that with no DNA input (panels B and C) at 12 hours of amplification, non-specific products are observed on the gel. The source of these is the very small amount of DNA contamination present in the random hexamers and other reagents of the TempliPhi kit. Because of the extreme sensitivity of the method, this contaminating DNA is amplified in the absence of circular DNA input at longer reaction times. However, an addition of plasmid or other circular DNA to the reaction completely suppresses this background amplification (Figure 7.4 B, C).

Chapter 7. Plasmid Preparation Methods

Additional bands visible in the wells and below the main restriction band may represent aberrant, non-cleaved forms of DNA produced during RCA and do not interfere with subsequent analyses. Figure 7.3C also shows the results of TempliPhi reactions carried out on bacterial colonies containing pUC18.

Plasmid DNA amplified with TempliPhi may be sequenced directly by cycle sequencing. Figure 7.5 shows the sequence of pUC18 DNA obtained by cycle sequencing on a sample isolated from a single colony after amplification with TempliPhi. Excellent quality sequence with a PHRED Q > 20 value of 650 bases is consistently obtained (L. Hosta, Amersham Biosciences; personal communication). Likewise, larger cloning vectors such as cosmids and BACs may be amplified and sequenced starting with partially purified DNA obtained from mini-lysates or colonies. A modified TempliPhi protocol has been published for BAC amplification and sequencing (1). It is recommended that at least 10 ng of starting BAC DNA be used as input, an amount that can easily be obtained from an overnight 1.5 ml culture followed by a standard mini-alkaline lysis protocol. This amount of input DNA routinely yields 10 to 15 µg of amplified BAC DNA, of which 2 µg is used for each sequencing reaction. Sequence quality and pass rates are equivalent to those of the best current methods for BAC DNA preparation (1).

Comparison of TempliPhi to Methods Based on Culturing

It is instructive to perform a comparison of TempliPhi to other plasmid preparation methods in terms of performance and cost. Table 7.2 presents a cost comparison of the most popular products currently on the market. However, a number of factors need to be taken into account in order to make fair comparisons. First, it is important to restate that TempliPhi does not require overnight culturing, yielding considerable time savings. Also, users of the standard alkaline lysis-based methods need to spend additional money on labor and consumables. An estimate of the maximum cost for consumables (plastic tubes, tips, media, etc.) with these methods is $0.15/prep. There are additional costs associated with running and maintaining instruments (centrifuges, etc.), estimated to be $0.50/prep maximum. Hence, although the cost per prep for TempliPhi based strictly on the kit cost falls above the middle of the range for plasmid preparation kits, other costs for consumables, reagents, waste disposal, additional equipment (vacuum manifold, magnetic plate), and overhead and the additional time required for culturing make TempliPhi much more cost-competitive. Also, it should be pointed out that the volumes of plasmid preparations often can be scaled down considerably from the supplier's recommendations, significantly reducing the cost per prep (16).

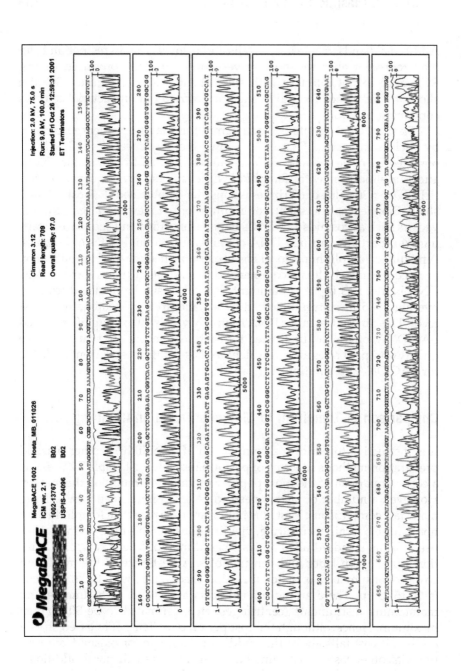

The principal benefits of TempliPhi over alkaline lysis-based culturing methods can be summarized as follows:

- Eliminates the need for culturing bacteria saving at least one day of prep time.
- Plasmid amplification is complete in only four hours yielding 1 to 1.5 µg of quality template for sequencing or other applications.
- No need for individual protocols for plasmid, cosmid, fosmid, BAC, and phage preparation.
- Amplification is isothermal; thus, there is no need for thermal cyclers.
- Amplification does not require custom primers.
- Low or high copy-number plasmids can be amplified to yield a similar amount of DNA.
- Amplified DNA is used directly for sequencing without further purification.
- Multiple reactions easily can be performed at the same time, resulting in very high throughput.

Summary

Methods for plasmid DNA preparation have evolved to make the process shorter, more efficient, and more automatable. Most currently available commercial products are based on a modified alkaline lysis protocol followed, after debris clearance, by selective binding of plasmid DNA to a silica matrix in the presence of an appropriate salt. After washing, the DNA is eluted with water or low salt buffer and either used as is or concentrated further by alcohol precipitation. Some suppliers have introduced paramagnetic beads for binding the plasmid DNA, eliminating the need for centrifugation or vacuum elution steps. A magnetic plate or disc is used to move the beads to the bottom of the tube or well. Most of these methods can be carried out in microtiter plates for high-throughput applications and may be automated with appropriate robotics.

Figure 7.5. **MegaBACE 1000 sequencing results of a plasmid amplified from a bacterial colony using TempliPhi.** A portion of a single bacterial colony containing pUC18 plasmid DNA was picked with a pipette tip and added to 5.0 µl of TempliPhi Sample Buffer. The sample was incubated for 8 hours after the addition of 5.0 µl of a mixture containing TempliPhi Reaction Buffer and Enzyme. A volume of 1.0 µl was then sequenced with the DYEnamic ET Terminator Cycle Sequencing Kit (Amersham Biosciences) using the recommended protocol. Electrophoresis was carried out on the MegaBACE 1000 instrument (Amersham Biosciences). The same quality of sequence data has been obtained on the ABI 3700 and 3100 instruments (Applied Biosystems) (data are from Amersham Biosciences and J. Kieleczawa, personal communication; data not shown).

Table 7.2. Cost comparison of plasmid preparation products from various suppliers.

Supplier	Product	Format	Pack Size	Price/Prep[a] US$
Amersham Biosciences (www.amershambiosciences.com)	TempliPhi™	RCA on pg quantities of starting plasmid from cultures or colonies	100 reactions	$2.75
			500 reactions	$2.60
Agencourt Biosciences	MCPrep™ 96 Well Plasmid Purification Kit	96-well SPRI™ magnetic beads for binding DNA-up to 2 μg per well (NOTE: SPRI™ 96 magnetic plate sold separately —$500)	4,000 preps	$0.38
			35,000 preps	$0.35
Bio-Rad Laboratories	Quantum Prep®	Spin Columns— >20 μg DNA	100 preps	$1.04
Edge Biosystems	Plasmid 96 Miniprep Kit	96-well filter plate-centrifuge	10 plates —960 preps	$1.03
Eppendorf/Brinkmann	Perfectprep® Plasmid Miniprep Kit	Spin column (5–30 μg DNA)	250 preps	$0.92
Eppendorf/Brinkmann	Perfectprep® Plasmid 96 Spin Direct Bind Kit	96-well filter plate-centrifugation	384 preps	$1.50
Invitrogen/Life Technologies	Concert™ Rapid Plasmid Miniprep System	Spin Columns— up 20 μg DNA	50 preps	$1.16
Invitrogen/Life Technologies	Concert™ High Purity Plasmid Miniprep System	Anion exchange column—up to 20 μg DNA	25 preps	$4.60
Millipore	Montage™ Plasmid Miniprep 96 Kit	96-well filtration-vacuum	384 preps	$1.11
			2,304 preps	$0.95
Promega	Wizard® Plus Minipreps	Spin column-1–10 μg DNA	100 preps	$1.17

Table 7.2. *Continued*

Supplier	Product	Format	Pack Size	Price/Prep[a] US$
	DNA Purification System			
	Wizard® SV96 Plasmid DNA Purification System	96-well DNA binding plate Vacuum elution	480 preps	$1.57
	Wizard® Magnesil™ Plasmid Purification System	96-well paramagnetic particles for plasmid capture (NOTE: Magnabot 96 magnetic device sold separately—$478)	768 preps	$0.60
Qiagen	Qiagen Spin Columns	Individual columns-20 µg per column	250 columns (preps)	$0.98
	QIAprep™ 96 Spin Miniprep Kit	Spin columns—up to 20 µg	250 preps	$1.00
	QIAprep™ 96 Turbo Miniprep Kit	96-well plates (filter and capture) Vacuum elution (NOTE: QIAvac manifold sold separately—$402) —up to 20 µg	384 preps	$1.90
	QIAwell™ 96 Ultra Plasmid Purification Kit (high purity, high throughput)	96-well additional plate step between filter and prep plate-up to 20 µg	384 preps	$4.36
Stratagene	StrataPrep®	Spin column-up to 20 µg DNA	250 preps	$0.98
Whatman	96-Well Plasmid Miniprep Kit	96-Well DNA binding plate	384 preps	$1.56

[a] Based on catalog list price.

A novel DNA amplification product, the TempliPhi Kit from Amersham Biosciences, was recently introduced for plasmid preparation without bacterial culturing. Starting with as little as picogram quantities of plasmid DNA from liquid cultures or single colonies, TempliPhi yields microgram quantities of product in as little as four hours. The amplification is isothermal (30°C) and the product can be used directly in DNA sequencing, restriction analysis, and other applications. Because of the large savings in time and consumables achievable with TempliPhi, this product has revolutionized the field of plasmid preparation for sequencing.

References

1. Amersham Biosciences. 2002. Improved method of BAC DNA preparation for sequencing. Application note 63-0048-57. Available at: http://www.amersham.com. Accessed January 2004.
2. Birnboim, H.C., and Doly, J. 1979. A rapid alkaline extraction procedure for screening recombinant plasmid DNA. *Nucleic Acids Res* 7: 1513–1523.
3. Birnboim, H.C. 1983. A rapid alkaline extraction method for the isolation of plasmid DNA. *Methods Enzymol* 100: 243–255.
4. Blanco, L., and Salas, M. 1984. Characterization and purification of a phage phi29-encoded DNA polymerase required for the initiation of replication. *Proc Natl Acad Sci USA* 81: 5325–5329.
5. Blanco, L., and Salas, M. 1985. Characterization of a 3'–5' exonuclease activity in the phage phi29-encoded DNA polymerase. *Nucleic Acids Res* 13: 1239–1249.
6. Blanco, L., Bernad, A., Lazaro, J.M., et al. 1989. Highly efficient DNA synthesis by the phage phi29 DNA polymerase. Symmetrical mode of DNA replication. *J Biol Chem* 264: 8935–8940.
7. Dean, R.B., Nelson, J.R., Giesler, T.L., and Lasken, R.S. 2001. Rapid amplification of plasmid and phage DNA using phi29 DNA polymerase and multiply-primed rolling circle amplification. Genome Res 11: 1095–1099.
8. Dean, F.B., Hosono, S., Fang, L., et al. 2002. Comprehensive human genome amplification using multiple displacement amplification. *Proc Natl Acad Sci USA* 99: 5261–5266.
9. Garmendia, C., Bernad, A., Esteban, J.A., et al. 1992. The bacteriophage phi29 DNA polymerase, a proofreading enzyme. *J Biol Chem* 267: 2594–2599.
10. Hawkins, T.L., O'Connor-Morin, T., Roy, A., and Santillan, C. 1994. DNA purification and isolation using a solid-phase. *Nucleic Acids Res* 22: 4543–4544.
11. Lasken, R.S., Dean, F.B., and Nelson, J.R. 2001. Multiply-primed amplification of nucleic acid sequences. U.S. Patent No. 6,323,009.
12. Malek, J.A., McKernan, K.J., and McEwan, P.J. 2001. Automating BAC ends sequencing for large-scale genome projects. XIIIth Genome Sequencing and Analysis Conference. San Diego, CA, Oct. 25–28, 2001. 28 (abstract).

13. Nelson, J.R., Cai, Y.C., Giesler, T.L., et al. 2002. TempliPhi™: Phi29 DNA polymerase-based rolling circle amplification of templates for DNA sequencing. *Biotechniques* 32: S44–S47.
14. Schwartz, J., Donaldson, R., and Davis, R. 2001. Implementation of a magnetic bead dye terminator cleanup protocol on a novel small volume thermalcycler. XIIIth Genome Sequencing and Analysis Conference. San Diego, CA, Oct. 25–28, 2001. 29 (abstract).
15. Shah, P.C., Dar, M. Dhulipala, R., et al. 2002. A family of amplification technologies for the post-genome era. 14th Genome Sequencing and Analysis Conference, Boston, MA, Oct. 2–5. F-25 (abstract).
16. Wu, H.-C., Shieh, J., Wright, D.J., and Azarani, A. 2003. DNA sequencing using rolling circle amplification and precision glass syringes in a high-throughput liquid handling system. *Biotechniques* 34: 204–207.

8 Optimization of Culture Growth in 96-Deep-Well Plates

Jan Kieleczawa
Wyeth Research, Cambridge, MA

Introduction

The growth of plasmid DNAs in 96-deep-well plates provides a convenient and efficient way to prepare a substantial number of DNA samples for library screening, mutant isolations or evaluation of experimental conditions, etc. The 8 × 12/96-well plate footprint is universally adapted for high-throughput robotic processing and is compatible with 8- and 12-channel manual pipettes. This compatibility allows manual preparation of 400 to 800 DNA plasmids/8-h day/person depending on the specific plasmid isolation method.

The amount of plasmid DNA needed depends on the type of experiment and may vary from a few nanograms (e.g., for PCR experiments) to a few micrograms (e.g., various restriction analyses, full-length double-stranded DNA sequencing). For full validation, most experiments need to be repeated a few times; hence, as a general rule, the more DNA obtained from one preparation, the better.

This chapter describes a number of simple parameters that can be optimized to get the maximum amount of plasmid DNA from cultures grown in 96-deep-well plates. Furthermore, we assume that most typical core laboratories do not use sophisticated culture growth chambers and the suggested improvements can be implemented without additional capital expenditures.

Materials and Methods

The compositions of growth media were as described in (1, 2). Most experiments described in this chapter involved DH5 alpha competent cells (Invitrogen, Inc., Carlsbad, CA), chemically transformed with pGem3zf DNA (purchased from Promega, Madison, WI), or other plasmids, grown on a standard LB agar plates with 20 μg/ml of ampicillin or 12.5 μg/ml of kanamycin. Unless otherwise stated, the experiments were performed as follows: colonies incubated overnight were picked from an agar plate and inoculated into 1 ml of an appropriate medium in 96 deep well plates (Qiagen, Inc., Valencia, CA). The 96-well plates, covered with lids, were placed (time = 0 hours) in an air shaker (model 25D; New Brunswick Scientific, New Brunswick, NJ) set at 37°C and in most experiments the speed was maintained at 350 rpm. At specified intervals, 10 μl aliquots were withdrawn and transferred to 90 μl of water for $OD_{600\,nm}$ measurements in a Perkin-Elmer Lambda 2S spectrophotometer (Perkin-Elmer, Norwalk, CT). For comparison, the growth of some plasmid DNAs was followed in 1 ml cultures placed in standard 10 ml glass tubes under the same conditions. Each data point is an average of three to five individual measurements for a given series. If the experiment involved the DNA plasmid preparation, the 96-Well Alkaline Lysis Miniprep Kit was used (EdgeBiosytems, Inc., Gaithersburg, MD) and 3 μl aliquots (out of total of a 100 μl) were run on 1% agarose gel/1 × TAE buffer as described in (2). For quantification, a dilution series (25–500 ng) of pGem3zf was loaded on each agarose gel and the DNA concentrations were estimated from Polaroid pictures (Polaroid Corp., Cambridge, MA).

Experimental Parameters Optimized

The following basic growth parameters were evaluated: the type of growth media, the speed of shaking (rpm), the length of growth, the volume of growth cultures, and the concentration of ampicillin.

1. Growth media and shaking speed. To get the maximum amount of plasmid DNA, we tested three rich-growth media: Terrific Broth (TBr), 2x YT, and M9TBY. In this experiment, in addition to a standard pGem3zf plasmid (ampicillin resistance), which is a high copy number plasmid (1, 2), a pACYC177 plasmid (New England BioLabs, Beverly, MA) was used. The pACYC177 is a medium copy number plasmid with kanamycin and ampicillin resistance markers (1, 2). Table 8.1 shows the data for these two clones grown for 24 hours at 37°C at three shaking speeds. Two general conclusions can be drawn from these data. First, the cells grown in Terrific Broth medium have the highest optical density, which is proportional to the amount of cells (1, 2) and, in general, to the amount of DNA that can be extracted. Second, the amount of cells increases with higher shaking speed: at 350 rpm the number of cells doubled

Table 8.1. Relative growth of plasmid DNAs at different shaking speeds.

Clone	Media	$OD_{600\,nm}$		
		250 rpm	300 rpm	350 rpm
pGem3zf	TBr	5.3 ± 0.5	6.6 ± 0.7	9.3 ± 0.7
	2xYT	4.3 ± 0.4	5.8 ± 0.5	8.5 ± 0.6
	M9TBY	4.0 ± 0.5	5.3 ± 0.7	7.9 ± 0.8
pACYC177	TBr	2.4 ± 0.3	3.5 ± 0.6	4.8 ± 0.5
	2xYT	2.0 ± 0.5	3.0 ± 0.6	4.5 ± 0.4
	M9TBY	1.7 ± 0.4	2.9 ± 0.5	3.8 ± 0.5

Five 1-ml cultures were incubated in different growth media and at the indicated shaking speed. After 24 hours the $OD600_{nm}$ was measured for each separate sample as described in the **Materials and Methods** section.

compared to cultures grown at 250 rpm. The shaking speed limit of 350 rpm was due to the type and condition of the shaker used in this experiment, and for different shakers the speed can be higher.

2. Length of growth. Under optimal nutrient conditions, most bacteria will follow the classical four-stage growth pattern: lag phase, exponential growth, steady-state phase, and the decline phase (1, 2). To evaluate the time necessary to obtain the optimal amount of plasmid DNA pGem3zf in DH5 alpha was grown as described in the **Materials and Methods** section. At specified times three independent 1 ml samples were withdrawn, and spun down and pellets were frozen till the end of a series. Upon completion of the time-course experiment, plasmids were separately prepared from 1 ml cultures, as descried above, and aliquot DNAs were run on an agarose gel. Figure 8.1 shows the total amount of recovered pGem3zf DNA at various time points. This and many other experiments with different insert sizes (not shown) suggest that 20 to 24 hours of growth is sufficient to obtain the optimal amount of DNA for high-copy number plasmids such as the pGem series. The total amount of plasmid recovered under these conditions varies from 2 to 15 µg/ml of culture depending on the type and size of insert.

3. The volume of growth culture. Under the experimental conditions described above, it is possible to grow cultures in up to 1.75 ml without cross-contamination (not shown). This growth volume, however, is not optimal for the recovery of plasmid DNA. Figure 8.2 shows the total amount of recovered DNA from different culture volumes. It appears that under these conditions, the optimal culture

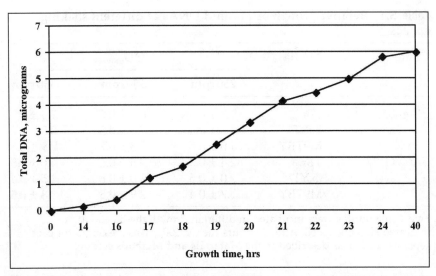

Figure 8.1. The total amount of plasmid DNA recovered at different growth times. At each data point three separate 1 ml aliquots were withdrawn and culture centrifuged for 5 minutes at 14,000 × g. Pellets were frozen until all data points were collected and processed. The total amount of DNA was estimated from agarose gels as described in the **Materials and Methods** section. The cultures were grown at 37°C/350 rpm.

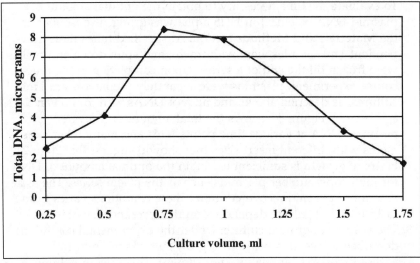

Figure 8.2. The total amount of plasmid DNA recovered from different volumes of culture. Pre-incubated pGem3zf culture was distributed in triplicate into a 96-well plate and grown for 24 hours at 37°C/350 rpm. The total amount of recovered plasmid DNA was estimated as described in Figure 8.1.

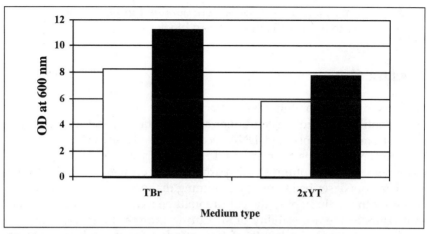

Figure 8.3. Comparison of bacterial cell growth in a 96-well plate and glass tubes. Three 1-ml cultures of pGem3zf were grown in two different media either in a 96-well plate (□) or in 10-ml glass tubes (■). After 24 hours (37°C/350 rpm) aliquots were withdrawn and the OD measured at 600 nm.

volume is between 0.75 and 1 ml. At lower or higher culture volumes the limiting factors are the total volume and the sufficient aeration, respectively.

4. The amount of antibiotic. In this experiment, eleven different clones from a library of *Borrelia burgdorferii* (insert sizes 1–3 kbp) and pGem3zf control plasmid were grown for 40 hours in the presence of eight different concentrations (0, 20, 40, 60, 80, 100, 200, and 400 µg/ml) of ampicillin. On average, the amount of recovered DNA is slightly higher at ampicillin concentrations above 80 µg/ml (not shown).

5. Growth in 96-deep-well plates versus growth in glass tubes. Assuming that nutrients are not a limiting factor, the critical parameter in effective growth of bacterial cultures is sufficient aeration. In some cases, sophisticated and expensive equipment (e.g., HiGro shakers from GeneMachines, San Carlos, CA) can be used to grow cultures under optimal air conditions. As shown in Table 8.1, simply increasing shaking speed by 100 rpm will double the amount of DNA recovered from 1 ml cultures grown in 96-deep-well blocks. The other parameter that influences the amount of air available to cells is the surface area of a growth well or a tube. The ratio of surface area over the culture in a typical 10-ml tube to an individual well in a 96-deep-well block is approximately 1.5. Figure 8.3 shows the culture growth (pGem3zf/DH5 alpha/350 rpm/37°C) in two different media in 96 deep wells and in 10-ml glass tubes.

Approximately 40% more cells were obtained in glass tubes compared to 2-ml wells in a 96-deep block.

Conclusions

There are limitless combinations of vectors, type and size of DNA inserts, media, antibiotics, growth conditions and other factors that will affect the optimal amount of plasmid DNA recovered from a liquid culture. The data presented in this chapter should serve only as a general guide for those who wish to optimize their particular experimental conditions. Of all the factors described here, and assuming that one selects the most suitable growth medium, only the culture volume may be a universal value, as it refers to the accessibility of air during growth under these experimental conditions. The length of time needed for the culture to reach optimal saturation depends on many factors, including size, type of insert, and type of vector (especially whether it is high, medium or low copy number). From the experiments presented here, it appears that incubation for 20 to 24 hours is needed for optimal growth with various inserts in high-copy number vectors in 96-well plates. When 1-ml cultures were grown in 10-ml glass tubes, cell densities were approximately 40% higher (Figure 8.3). When bacterial cell cultures were grown under more standard growth conditions (e.g., 100-ml erlenmeyer flask with 20-ml culture) the time needed for full cell saturation was about 10 to 14 hours (data not shown).

References

1. Ausubel, F.M., Brent, R., Kingston, R.E., et al. (eds). 1998. *Current Protocols in Molecular Biology*. John Wiley & Sons, Inc. New York.
2. Sambrook, J., and Russell, D.W. 2001. *Molecular Cloning*, 3rd ed. Cold Spring Harbor Laboratory Press. Cold Spring Harbor, NY.

9 Automated DNA Scanners Used in Sequencing Laboratories

Jan Kieleczawa
Wyeth Research, Cambridge, MA

Introduction

The introduction of the first fluorescence-based DNA sequencer (3, 4) made it possible to electronically gather, store, and manipulate sequence data. Continuous technological improvements together with more robust chemistries led to the development of almost fully automated capillary-based DNA sequencing systems, which made the sequencing of many bacterial genomes and eventually the human genome possible (1, 5). At present, a number of low-, medium-, and high-throughput sequencing machines exist on the market (ABI 3700, 3100 and 3730; MegaBACE 1000 and 4000; SpectruMedix 96 and 192; Beckman CEQ2000/8000) that can fulfill the needs of any laboratory. Recently, both Applied Biosystems (Foster City, CA) and Amersham Biosciences (Piscataway, NJ) introduced more personalized versions of their respective machines (ABI3100, ABI3730 and MegaBACE500) to provide more flexibility and savings in reagents, and to increase the read length of sequencing runs, which is very valuable in closing remaining gaps in genomes of interest. We can expect that in the near future even more personalized (benchtop-like) DNA sequencers will be available to make data gathering quicker, friendlier, and cheaper. For example, a number of academic labs and commercial companies are working on miniaturization of DNA sequencing devices that hold the promise to deliver several hundred bases/read in just a few minutes (see also Chapter 13). Such devices will be invaluable for quick checks, DNA sequence confirmations, etc., both in the laboratory setting and in the field.

Brief History of Some Automated Scanners

This section provides an overview of major milestones in the development of automated DNA sequencers. Since the very beginning the leader in the field was (and still is) Applied Biosystems. Amersham Biosciences holds the distant second position in the market share, though it is credited with the development of the first multicapillary DNA sequencer. Other manufacturers, Beckman-Coulter, Spectrumedix, Li-COR and MJ Research, have developed sequencing instruments that may have some technological innovations (e.g., in-capillary detection, infrared dye-terminators) but failed to capture a substantial market share. Data on these instruments is provided in Table 9.1. Most DNA sequencing instruments also are capable of performing genotyping using the same system configuration. Some of the data presented here were obtained from the manuals accompanying sequencing instruments and are not referenced.

1987, ABI370—the first automated DNA sequencer based on four-color technology. It had a 16 lane capacity and PMT detection; slab gel technology.
1989, ABI373—36 lanes, PMT detection, gel slab technology.
1993, ABI373XL—64 lanes, PMT detection, slab gel technology. Both the ABI373 and 373XL DNA sequencers are very reliable and quite user-friendly machines. Throughout most of the 90s (and in many cases still today) they were true workhorses in many academic labs, core centers, and commercial units. By today's standards they are obsolete and their technical support by Applied Biosystems was discontinued after January 1, 2003.
1995, ABI377—96 lanes, CCD detection, slab gel technology. This was the first truly high-throughput DNA sequencer with the potential to run up to 384 samples/day. Easy operation and CCD detection made it very flexible and able to accommodate a number of different chemistries. These reasons (and the fact that currently they are relatively inexpensive on the second-hand market) assure that the ABI377 still enjoys strong market presence (30% of the market) (2). We hope that Applied Biosystems will support it for several more years. No ABI slab gel instrument is in production any longer.
1995, ABI310—this was the first single-capillary DNA sequencing instrument on the market. It is still in production and used in many low throughput laboratories.
1997, MegaBACE 1000—This was the first multicapillary DNA sequencing instrument on the market. It performs very well with respect to read lengths and quality of data; however, it is quite sensitive to variability in DNA concentration. All MegaBACE instruments feature user-friendly and versatile software managed through the Instrument

Chapter 9. Automated DNA Scanners

Control Manger (ICM). Its Sequence Analyzer module is phred-compatible for easy downstream integration.

1999, ABI3700—The ABI3700 was the first truly semiautomatic, multiplate, high-capacity instrument that made the sequence of the human genome and other organisms possible. It remains a staple in many DNA sequencing factories. No longer in production, its support is being phased out.

2001, MegaBACE4000—This instrument is a high-throughput version of MegaBACE1000 with fourfold increase in sample processing capacity. It performs simultaneous injection and runs up to 384 samples.

2001, MegaBACE500—This model is especially targeted for core facilities and smaller DNA sequencing labs. If needed, it can be easily upgraded to a MegaBACE1000 model for higher throughput.

2001, ABI3100—This instrument became an instant favorite in many core and forensic laboratories. They are very easy to operate, cost effective and produce data of very high quality. Depending on the user's requirements and specific application, the machine can be equipped with 36-, 50- or 80-cm long capillary arrays. When equipped with an 80-cm capillary array, this machine is an excellent tool in DNA finishing projects with read lengths exceeding 1100 bases. Compared to the ABI3700, it has more elegant and economic delivery of polymers and buffers, resulting in substantial cost savings.

2002, ABI3730/3730*xl*—The ABI3730 model is a scale up of the ABI3100 with 48 capillaries loaded and run simultaneously. The ABI3730*xl* is the 96-well, high-end version of the ABI3730, almost fully automated and primarily designed for DNA sequencing factories. Both these models have technical solutions similar to the ABI3100. ABI3100/3730 and 3730*xl* were developed in collaboration with Hitachi, Ltd. (Tokyo, Japan).

DNA Sequencing Instruments—Product Overview

Table 9.1 provides more detailed characteristics of major DNA sequencing instruments. Data for all ABI units are based on this author's experience and company literature. Data for most other sequencers are mainly from the respective company sources.

Criteria for Selecting a DNA Sequencing Instrument

The choice of a DNA sequencing instrument is primarily based on the nature of DNA sequencing work in any given laboratory, required throughput, prior experience with a specific brand, and to some degree the outlook for present and future market conditions. Currently two

Table 9.1. Characteristics of main DNA sequencing instruments.

DNA sequencer—number of caps	Read length (bases)	Accuracy (%)	Possible daily throughput (kpbs)	Cost per reaction ($)	Separation method
ABI[a] 377-96	500–700		135,000	1–2[a]	Slab gel
3700-96	500–900		360,000	1–2[a]	Capillary
3100-16	900–1100	98.5–99.0	82,000	1–2[a]	Capillary
3730-48	500–1100		864,000	<1[c]	Capillary
3730x l-96	500–1100		1,728,000	<1[c]	Capillary
AB[a] MB 500-48	600–750		390,000	<1[c]	Capillary
MB1000-96	600–750	98.5	790,000	<1[c]	Capillary
MB4000-384	600–650		2,800,000	<1[c]	Capillary
SM[a] SCE2410-24	500–800		168,000	<1[c]	Capillary
SCE9610-96	500–800	99.3	672,000	<1[c]	Capillary
SCE192-192	500–800		1,344,000	<1[c]	Capillary
LI-COR 4300S	800	99.0	102,000	1–5[b]	Slab gel
4300L	1100	99.0	140,000	1–5[b]	Slab gel
BC[a] CEQ2000	600–800	98.5	67,000	1–2[c]	Capillary
CEQ8000	600–900	98.0	82,000	1–2[c]	Capillary
MJ[a] BST51	500–1100	98.5	230,000	1.56[b]	Both: Capillary
BST100	500–1100	98.5	480,000	1.56[b]	gel

The data in this table were compiled from marketing brochures, technical bulletins, and company Web sites. Listed prices for sequencing instruments are as quoted by the respective companies at the end of 2003. They may be substantially different, depending on various factors such as the number of purchased instruments, trade-ins, the need to break into a specific market, etc. Cost per reaction includes only the cost of reagents, as other cost factors, mainly overhead and amortization, vary in different institutions.
[a] Cost based on data from low-throughput DNA sequencing laboratories.
[b] Cost based on data from a manufacturer.
[c] Cost estimated from costs of reagents. It is important to note that cost/reaction may vary significantly between DNA sequencing centers and needs to be calculated using the most accurate site-specific data.
[d] The ABI377 is no longer sold by Applied Biosystems and may be purchased on secondary market. Price may vary.

	Characteristics			
Detection method	Strengths	Weaknesses	Company's phone and Web site	Price/ unit in 2003 ($)
CCD	Reliable	Labor intensive		~10,000[d]
CCD	Fully automated	"Buggy" software	800-831-6844	300,000
CCD	Long reads	Frequent loss of 1st gel	appliedbiosystems.com	136,000
CCD	Lower usage of buffers & POP	N/A yet		250,000
CCD		N/A yet		350,000
Scanning confocal microscope	Scalability	Lack of automation. Sensitive to template concentration	800-526-3593 amersham biosciences.com	200,000
	Throughput			350,000
On-column CCD camera	All: walk-away automation and high sensitivity	All: small company's size. Will they be here tomorrow?	814-867-8600 spectrumedix.com	130,000 250,000 350,000
Scanning fluoresc. microscope	Both: Accuracy and longest reported reads	Both: Relatively low throughput and cost/ reaction more expensive than others	800-645-4267 licor.com	50,000 57,000
Four filters wheel and PMT	Both: integrated downstream applications	Both: Low throughput but fills in an important niche	800-742-2345 beckmancoulter.com	100,000 120,000
Both: 4 PMT	Both: fast run times and good analysis software	Both: need to prepare gels makes it labor intensive	888-735-8437 mjr.com	81,500 131,000

producers, Applied Biosystems (ABI) and Amersham Biosciences (AB) hold close to 95% of the market for automated DNA sequencers (ABI about 80% and AB about 15%, respectively; USB Warburg survey). The remaining manufacturers hold the remaining 5% of the market (the exact distribution held by of these five manufacturers is hard to estimate). Applied Biosystems holds almost monopolistic dominance of DNA sequencers market and choice of a machine from ABI that suits individual needs is a relatively safe bet. This is especially comforting (in this author's view) as the customer and technical support service by ABI personnel has improved substantially over the past few years. Due to the breadth and depth of choices from ABI in DNA sequencing instruments, any current need can be met by one of their machines. For core or small-throughput laboratories the choice of the ABI3100 is particularly attractive, as this instrument can be equipped with 36-, 50-, or 80-cm long capillary arrays. One of the solutions, if possible, is to have two instruments equipped, for example, with 50- and 80-cm capillaries to accommodate different needs for read length. The recently introduced ABI3730 offers higher throughput (48 capillaries) as well as improved design for substantial cost savings in polymer, buffer, and enzyme mix. In addition, its modular design provides an option to upgrade this instrument to the full production version (ABI3730*xl*), if needed. The company's data estimate that the ABI 3730/3730*xl* consumes 30 to 40 times less buffer and polymer, and generates up to 90% less waste compared to the ABI3700.

Ease of use and maintenance, as well as data quality/read length/run times are of critical importance when making a purchasing decision. While the ABI3700 produced good quality data with phred scores of 550 to 700 Q ≥ 20 bases, its maintenance requires considerable technical expertise. Periodic change of the capillary array takes up to one day, which may cause disruption in DNA sequencing service. On the other hand, the ABI3100 is extremely easy to maintain: capillary change or re-installation and calibration take just a few minutes and neither requires any extensive technical skills. This is especially important if the instrument is not used for a prolonged period of time (over a week) when it is recommended to store the capillary array off the machine. An ABI3100 equipped with a standard 50-cm capillary produces good quality data with phred scores in the range of 650 to 850 Q ≥ 20 bases, and it is possible to run up to 144 samples/day. An ABI3100 equipped with an 80-cm capillary array is capable of running 96 reactions/day and producing data with phred scores in the range of 950 to 1100 Q ≥ 20 bases. Finally, the ABI3730 and 3730*xl* are capable of completing up to 12 long runs/day (576 or 1152 samples) with quality scores in the range of 800 to 950 Q ≥ 20 bases. The other important feature of recent ABI (and other manufacturers) instruments is improved software applications. Redesigned base callers produce longer read lengths, for example, ABI's KB-caller offers on average 100 bases more data than the standard base caller. Standard software packages, which come with sequencing instruments, incorporate functionality that associates each called base with its phred or phred-like quality value and makes it easier for small laboratories to use them in their daily operations.

In years to come, capillary-based instruments will remain the basis for DNA sequencing services in most small centers as they produce long, quality reads. Hopefully, with further technological advances (instrumentation, polymers and enzyme mixes), quality read lengths will increase at the same time as the overall cost of sequencing steadily decreases. In Chapter 13, several, noncapillary-based solutions for DNA sequencing are described. Most of these designs, however, are still in the research and development stages.

Acknowledgments

I would like to thank Kathy Lee and Marianne Hane of Applied Biosystems for providing the data on older ABI sequencing instruments.

References

1. International Human Genome Sequencing Consortium. 2001. Initial sequencing and analysis of the human genome. *Nature* 409: 860–921.
2. Khanna, V., and Harper, M. 2002. *DNA Sequencer User Survey*. UBS Warburg. New York. Available at: www.ubswarburg.com/researchweb. Accessed January 2004.
3. Prober, J.M., Trainor, G.L., Dam, R.J., et al. 1987. A system for rapid DNA sequencing with fluorescent chain-terminating dideoxynucleotides. *Science* 238: 336–341.
4. Smith, L.M., Sanders, J.Z., Kaiser, R.J., et al. 1986. Fluorescence detection in automated DNA sequence analysis. *Nature* 321: 674–679.
5. Venter, J.C., Adams, M.D., Myers, E.W., et al. 2001. The sequence of human genome. *Science* 291: 1304–1351.

10. Geospiza's Finch-Server: A Complete Data Management System for DNA Sequencing

Sandra Porter, Joe Slagel, and Todd Smith
Geospiza, Inc., Seattle, WA

Introduction

The increased demands for DNA sequencing services and greater volume of data have created a need for software designed to support sequencing activities and applications. Commercial data management systems, such as the Finch-Server (Geospiza Inc., Seattle, WA), provide laboratories with robust data processing and quality analysis tools designed specifically to support DNA sequencing. Even laboratories new to sequencing can implement these systems quickly to analyze projects and identify problem areas.

Data management systems offer a variety of benefits. Two immediate benefits are a decrease in cost and a shorter time to complete sequencing projects. Small increases in quality or read length can have a large impact on the number of sequencing reactions required to complete a project. Because the overall cost of a sequencing project is determined primarily by the number of reads, project costs are lowered when data management tools are used to detect and correct problems in a timely fashion. A later benefit to researchers is the ability to link sequence data to sample history. It is important to know, for example, if a sequence was obtained from a cloned DNA fragment or from a clinical sample that might contain a heterogeneous mixture of sequences.

Finch-Servers were designed to solve common problems and meet the data management requirements shared by many laboratories. This approach benefits customers because multiple users participate in the design, review, and testing process. Additional benefits include a shorter time frame from purchase to use and a more robust and adaptable system, unlike custom systems that require a long development process and significant time investment on the part of the laboratory.

This chapter describes the software requirements of different types of sequencing laboratories and discusses how Finch-Servers help those laboratories meet their needs.

The Geospiza Finch-Server

A Finch-Server is a Web application with selected components of the Finch-Suite and a relational database management system (RDBMS), either Oracle (Redwood Shores, CA) or Solid (Mountain View, CA), installed on a computer with a UNIX or Linux operating system. Technicians and researchers use Web browsers such as Internet Explorer or Netscape to interact with the Finch-Server through a local intranet, or, in the case of geographically dispersed sites, over a secure Internet connection. Finch-Servers operate either as stand-alone systems or within a distributed computing environment.

Finch-Systems are Finch-Servers designed for different types of laboratories and different data processing requirements. Each Finch-System includes a Finch-Server along with selected components of the Finch-Suite, a set of integrated software modules designed to support data management and analysis. These include the: Sequencing Request Manager, Instrument Manager, Chromatogram Manager, Assembly Manager, Data Repository Manager, and Basic Local Alignment Sequence Tool (BLAST) Manager. Components of the Finch-Suite are organized within the Finch Core DNA Sequencing System (Figure 10.1), and the Finch Assembly and BLAST Systems (Figure 10.2). The Complete Finch DNA Sequencing System contains all of the Finch-Suite components.

Many laboratories need to store large numbers of data files and sequence assemblies. Customers have reported storing well over half a million chromatogram files in the Finch-Server and generating sequence assemblies from over 60,000 reads. These systems also provide a way for researchers to maintain their original, unprocessed data files, thus making it possible to apply new base-calling technology or other analysis tools to older data and data from varied sources. Storage media can be added to accommodate an increasing quantity of data over time. The Finch-Server, therefore, is able to provide a robust, scalable system, for storing chromatogram files, sequence assemblies, sequence databases, and BLAST results.

Much of the data stored in the Finch-Server are presented in tables that can be sorted by column. For example, one can easily sort reads by the number of vector clones by selecting the appropriate column heading. Customized data views also can be generated using data browsers that

Chapter 10. The Finch-Server Data Management System

quickly extract selected information. A laboratory manager might use the data browser to view all the sequencing runs performed with a certain instrument. A researcher setting up an assembly might select all of the reads from a genomic library that match *E. coli* and use the move icon to transfer them to a "trash" folder, thus keeping *E. coli* sequences out of the assembly.

Further, Finch-Servers store information in a relational database management system, allowing users to obtain selected information in response to an SQL (Structured Query Language) statement. The capacity to perform customized queries using SQL and obtain customized reports and information is a powerful tool for additional research, process management, and project oversight.

Data Management for Core Facilities

Managing Sequencing Requests

Core laboratories in academic institutions and biotechnology companies have a variety of general and customized requirements that software systems must be able to meet. First, the laboratory's customers need a convenient method for submitting work requests. This is accomplished in the Finch-Server through Web forms that allow customers to make requests using their own desktop computers. Core lab customers use the Finch-Sequencing Request Manager to enter experimental information for experiments performed on different scales. This information is stored in the database, allowing retrieval at a later date. Web forms, designed to accommodate tubes, 96- or 384-well plates, or batch modes, simplify data entry and sample naming (Figure 10.1A). A configurable, automated naming system assigns unique names to each sample. For example, samples that are used for sequence assembly must be named in a certain way to be compatible with the Phrap assembly program (6). The Sequencing Request Manager helps customers name samples appropriately with a minimal amount of extra work.

Sample Tracking

Service laboratories also require the ability to track samples through each stage of the sequencing process. Not only can laboratory personnel determine where samples are located in a queue, the Finch-Server allows customers to monitor the status of their own samples over the Internet, saving time spent on the phone. At the end of the sequencing process, the Finch-Server simplifies data delivery. Customers can either view (and store) their data in the Finch-Server or download the data to their desktop computer.

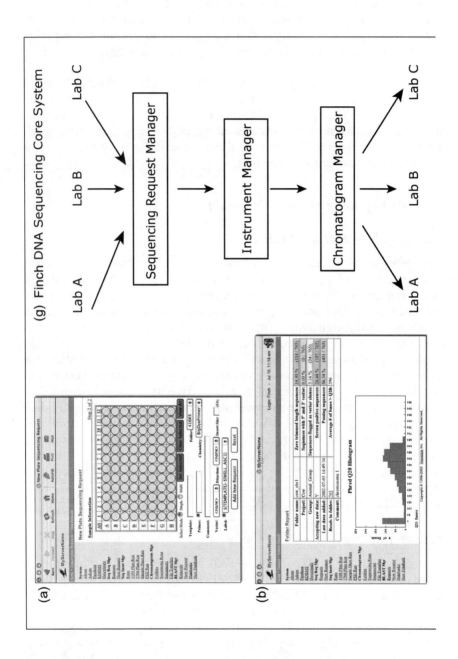

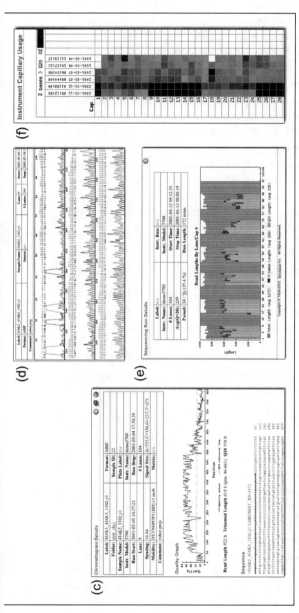

Figure 10.1. The Finch Core DNA Sequencing System. (a) A web page from the Finch-Sequencing Request Manager. Part of the form for submitting a request to sequence samples from a 96 well plate is shown. Single wells can be selected, or multiple wells, chosen by selecting column or row headings. (b) The Finch-Chromatogram Manager Folder Report. This report summarizes quality and selected statistics for all the reads in an individual folder, including the number of sequences that match vector sequences, *E. coli*, or other selected sequences. (c) The Chromatogram Details report, also from the Chromatogram Manager. Quality scores, from Phred, are shown for each base in the read, along with the sequence and other information. Links are available for viewing the trace file (d) and/or downloading the data. (d) A chromatogram trace file together with a plot of Phred quality scores. (e) The Sequencing Run Details report from the Finch-Instrument Manager. The read length (darker color) and sequence quality (light color) are shown on the y axis and the individual capillaries and/or lanes, on the x axis. The regularly repeating blocks of poor quality samples (indicated by a dark color) result from a plugged fang in the sequencing instrument. (f) The Instrument Capillary Usage report. Average sequence quality is represented by colors ranging from bright green (best) to black (poor). Run dates are shown at the top, from right to left. Problem capillaries (or samples) are shown by the black boxes in the first two columns. (g) The Finch Core DNA Sequencing System overview. Laboratories submit requests using the Sequencing Request Manager. The lab uses the Instrument Manager to set up sample sheets and downloads them to the sequencing instrument. Data are loaded into the Chromatogram Manager and the customers get the results.

Security

Core facilities often have clients from different laboratories, making it important to control access to data. Customers, on the other hand, want to share data with collaborators and other members of their laboratory, yet simultaneously prevent data access by competitors. These requirements are further complicated by the needs of the core facility. Personnel in the core facility must be able to view all of the data in order to monitor the quality of the laboratory's work. This problem is solved in the Finch-Server with a user hierarchy that allows administrators and technicians wider access than the facility's customers and assigns researchers to specific lab groups. All researchers within a lab group are able to view data belonging to that group but are prevented from viewing data owned by others. Only the facility administrator can add or delete researchers from a lab group by the facility director. As a result, researchers can share data through the Web in a protected manner with their collaborators and selected colleagues, while laboratory personnel can monitor data quality without compromising proprietary information.

Preparation of Sample Sheets and Managing Instrument-Related Data

After a customer has submitted a work request to a service laboratory, the laboratory needs to review requests and organize workflow. Technicians use the Finch Instrument Manager to create sample sheets in the Finch-Server by combining samples from different work requests to produce the optimum number of samples for each sequencing instrument. Finch-Servers support all of the widely used sequencing instruments, including those from ABI instruments (Applied Biosystems, Foster City, CA), the MegaBACE (Amersham, Piscataway, NJ), the Beckman CEQ2000 (Beckman, Fullerton, CA), and others. Laboratory technicians download sample sheets to their sequencing instrument and complete the sequencing run. Instrument-related service information and serial numbers also can be stored in the Instrument Manager.

Management of Chromatogram Files and Quality Control

The Finch Chromatogram Manager provides a useful tool for storing sequences and chromatogram-related information. Chromatogram files from a completed sequencing run are uploaded and stored in the Chromatogram Manager. The Chromatogram Manager also can be used to store individual files or packages of related chromatogram files that have been downloaded from public or private databases through the Internet.

Chromatogram file data are checked by the Finch-Server during the upload process to prevent accidental loading of duplicate

Chapter 10. The Finch-Server Data Management System

chromatograms. This feature allowed one customer to diagnose an unsuspected problem in their instrument software. The instrument software apparently created two files with different file names but identical chromatogram data. This bug was confirmed when the customer checked with the instrument vendor. Once in the Finch-Server, the chromatogram files enter a data processing pipeline. Some of the steps in this pipeline include: base-calling, quality measurement, and quality trimming with either Phred (3–6), TraceTuner (Paracel, Inc.), or the KB base caller (Applied Biosystems); identification of vector sequences with Cross Match (6); and the generation of graphical reports that summarize run information and statistics for sets of sequences (examples are shown in Figure 10.1b–d). Data quality analysis can be customized using pipelines to select the base-caller depending on the sequencing instrument, or laboratory.

Sequencing Instrument Performance

Instrument reports provide feedback to laboratories about the status of each run and the performance of each sequencing instrument. Blocked liquid transfer devices or problem capillaries can be quickly identified by viewing graphical reports (Figure 10.1, e, f). The ability to monitor capillary performance for changes in data quality over time helps laboratories determine when instruments require service. Service information and identification numbers are easily stored in the Finch-Server, providing a means to quickly view the service and repair history for any sequencing instrument.

Billing

Shared facilities need to monitor use and bill customers accordingly, in a timely fashion. Billing numbers can be stored in the Finch-Server when work requests are submitted. Structured Query Language (SQL) statements can be used in conjunction with billing numbers or other identification, to generate reports that define how resources are used, the amount of work performed, who requested the work, and the request date. These reports can be generated automatically, through custom services, and delivered to accounting departments.

Data Management for Large-Scale Sequencing

Laboratories engaged in genomic and production sequencing need additional features in a data management system. Production sequencing, expressed sequence tags (ESTs) clustering, single-nucleotide polymorphism (SNP) discovery, and genomic sequencing require the ability to assemble sequences; set up, maintain, and update sequence databases; pipelines can be designed specifically for automated data processing, filtering, and BLAST searches (1). These additional requirements can be met by stand-alone Finch systems, such as the Finch Assembly System and/or the Finch BLAST System; or with the Complete Finch DNA Sequencing System, which includes all of the components of the core system in addition to the Finch-Assembly Manager, the Finch-Data Repository Manager, and the Finch-BLAST Manager. An overview of these components is shown in Figure 10.2.

DNA Sequence Assembly

The Finch-Assembly Manager works with the Chromatogram Manager to assemble sets of reads into longer, contiguous sequences. Folders with specific sets of reads are created and organized in the Chromatogram Manager. To assemble sequences, one chooses the appropriate folder and assembly program, and presses a button to start the assembly. The Assembly Manager provides different options for sequence assembly, including Phrap (http://www.phrap.org; Green, International Human Genome Sequencing Consortium, 2001), a parallel version, SPS-Phrap (Southwest Parallel Software, Albuquerque, NM, USA), or both.

The Finch-Server's Relational Database Management System (RDBMS) stores all the parameters and details of the assembly, making it easy to repeat the assembly later with new or additional data. Sequence assemblies also can be treated as a type of experiment and performed with sets of sequences and different parameters. Becuase all of the assembly results and conditions are stored in the RDBMS, the optimum parameters are easily located for later work.

Phrap assembles reads by comparing pairs of sequences, locating overlapping regions, and using quality information from the base-caller to help reconstruct the original sequence. The assembly can be improved by incorporating additional information about each read. For example, Phrap qualities—those used to build the final contig—incorporate information about the relative orientation of each read, and the sequencing chemistry. If a base has been confirmed by an alternate method (sequencing the opposite strand or using a different type of chemistry), Phrap assigns a higher quality value to the base at that position. The highest quality bases are used to build a mosaic, which represents the sequence of the contig.

Information from each assembly is provided in tables and graphical reports. Tables include information about the number of reads in each contig, the location where each read aligns, and links to the read sequence,

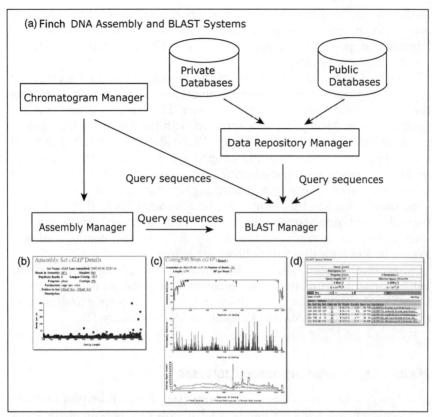

Figure 10.2. **The Finch-Assembly and BLAST Systems.** (a) The Finch-Assembly Manager uses read data, stored in the Chromatogram Manager, to set up sequence assemblies. The Finch-Data Repository Manager maintains current versions of public and private sequence databases. BLAST searches of selected databases are conducted with the Finch-BLAST Manager. Query sequences can be pasted in a form, uploaded from a file, or selected from either the Chromatogram Manager or the Assembly Manager. Data processing pipelines can be set up to perform queries in an automated fashion. (b) The Assembly Set Details report from the Assembly Manager. This report provides information from the assembly along with a graph showing the number of reads per kilobase (y axis) for different length contigs (x axis). (c) A contig report from the Assembly Manager shows the Phrap quality across the length of the contig (top), sequence discrepancies as a function of quality (middle), and the number of sequences mapping to each part of the contig (bottom). (d) A screenshot from the BLAST Manager shows BLAST parameters and a table with the results.

quality values, and trace file. Chimera reports are provided to help diagnose experimental artifacts. If further information is needed, say the locations of potential deletion clones, the Phrap output file can be downloaded for viewing.

Figure 10.2 b and c show two of the Assembly Manager's graphical reports. For this experiment, reads from several ESTs were obtained from the Washington University Genome Center (St. Louis, MO), stored in the Chromatogram Manager, and assembled with the Assembly Manager. A graph of the assembly results (Figure 10.2b) shows the number of reads per kb (y axis) versus the contig length (x axis). Because the reads used in this assembly were obtained from ESTs, sequences with a large number of reads/kb were likely to represent highly expressed genes, repetitive sequences, mitochondrial sequences, or ribosomal RNA. If these results were obtained from assembling reads from a genomic sequencing project, a large number of reads/kb might help flag problem contigs and find assembly problems due to repetitive sequences.

Graphical reports also provide an overview of the Phrap quality values for each contig, the coverage depth, and positions with discrepant bases. High-quality sequence discrepancies are a strong indicator of potential polymorphisms and valuable in SNP discovery (2).

Maintaining Updated Sequence Databases

Many companies and sequencing centers have found it helpful to maintain local copies of sequence databases. Not only can the searches be faster when using local resources, data are protected because the searches are performed in a secure environment. Researchers don't have to take risks with valuable information by sending proprietary sequences over the Internet. Further, local searches allow one to search proprietary or custom databases that aren't available over the Internet.

To meet this need, Geospiza developed the Finch BLAST System (Figure 10.2), which allows researchers to maintain up-to-date versions of any sequence database and use BLAST as a search tool. The content in public and private sequence databases changes on a daily basis, making it difficult for companies and research institutions to keep local versions current. The Finch-Data Repository Manager operates through a command line interface and updates local copies of databases on a regular schedule. Users specify which sequence data to retrieve from remote or local sites and designate where the files should be stored. When new sequences are retrieved, databases are automatically updated and log files are generated. E-mail reports instantly notify users that updates have occurred. This system also enables researchers to maintain multiple databases and multiple versions of each database.

The Finch-BLAST Manager

The Finch-BLAST Manager is used to perform BLAST searches of either nucleotide or protein sequence databases (Figure 10.2d). Researchers can search one or more databases, concurrently, with single or multiple sequences. Several of the BLAST programs are available in the BLAST Manager, including blastn, blastp, blastx, tblastx, and tblastn. The BLAST manager stores the search results and parameters, thus allowing comparisons to be made between database searches at different points of time.

Data processing pipelines can be employed by users to assemble reads stored in the Chromatogram Manager, and automated BLAST searches of selected databases performed with assembled sequences or individual reads. For example, a BLAST search can be carried out whenever updates occur in a specific database. BLAST can be employed also for quality control. In one example, a batch of clinical samples from polymerase chain reaction (PCR) assays was mixed up by a commercial sequencing lab. The problem was uncovered by constructing a custom database of the expected PCR products and comparing the reads with the expected products using BLAST.

Filters also are available in the BLAST Manager that allow users to mask low complexity sequences that might obscure search results. Repetitive sequences can be masked by adding RepeatMasker (Geospiza, Inc., Seattle, WA, USA) to the system. The addition of filters permits users to set up powerful screening systems designed to enhance data mining, genome sequencing, SNP discovery, or other applications.

Data Management Over the Internet: iFinch

iFinch is a Finch-Server designed for individual researchers and/or smaller laboratories with short term sequencing projects. Subscriptions to iFinch allow small laboratories immediate access to integrated data management tools without making a long-term investment. Unlike other Finch-Servers, Geospiza acts as the system administrator for iFinch, ensuring for example, that data are backed up on a regular basis, hardware is kept up to date, and that the system runs smoothly. Remote access allows laboratories to store data off-site, securely, minimizing the risk of data loss.

References

1. Altschul, S.F., Madden, T.L., Schäffer, A.A., et al. 1997. Gapped BLAST and PSI-BLAST: a new generation of protein database search programs. *Nucleic Acids Res* 25: 3389–3402.

2. Clifford R., Edmonson, M., Hu, Y., et al. 2000. Expression-based genetic/physical maps of single-nucleotide polymorphisms identified by the cancer genome anatomy project. *Genome Res* 8: 1259–1265.
3. Ewing, B., and Green, P. 1998. Base-calling of automated sequencer traces using Phred. II. Error probabilities. *Genome Res* 8: 186–194.
4. Ewing, B., Hillier, L., Wendl, M.C., and Green, P. 1998. Base-calling of automated sequencer traces using Phred. I. Accuracy assessment. *Genome Res* 8: 175–185.
5. Green, E. 2001. Strategies for the systematic sequencing of complex genomes. *Nature Rev Genet* 2: 573–583.
6. International Human Genome Sequencing Consortium. 2001. Initial sequencing and analysis of the human genome. *Nature* 409: 813–958.

11 DNA Sequencing Database: A Flexible LIMS for DNA Sequencing Analysis

Donald M. Koffman[1] and
Hemchand Sookdeo[2]
DMK Concepts, Inc.[1] *and Informed Solutions Corporation,*[1] *Winchester, and Wyeth Research,*[2] *Cambridge, MA*

Introduction

The DNA Sequencing Database is a cross platform (Wintel and Macintosh computers) Client Server database application, developed using the 4th Dimension database development system. 4th Dimension and 4D Server are products of 4D, Inc. (detailed in the company Web site: http://www.4d.com).

4th Dimension provides a modern integrated development environment for creating database management systems for small to intermediate size operations (of the order of 500 users). For larger systems, it can provide fully featured front ends to many databases (e.g., Oracle, Sybase, MS SQL Server, etc.). It also supports the development of a single application that can be compiled to run on a Wintel or a Macintosh platform as a stand-alone single user application or, using 4D Server/4D Client, as a multiuser application serving Wintel and Macintosh clients simultaneously.

DNA Sequencing Database is a multiuser application designed to serve both the scientists requiring DNA sequence determinations and the analysts providing the data. It not only captures information pertaining to Wyeth Research's Core Development DNA Sequencing Services, but also plays a major role in supporting, maintaining and facilitating the workflow of the analytical services department. Although it is primarily used to handle low-throughput sequencing analysis important for specific research programs, it has the flexibility to provide for high volume analysis as well, for example, end read clone surveys.

The DNA Sequencing Database supports the following workflow model of the operation:

Requestors (e.g., scientists) submit ("SEND") requests over the LAN network using a 4D client. Before requests can be submitted, the DNA Sequencing Database facilitates and enforces rules that insure that information required to perform the analysis is provided (e.g., DNA concentration, vector name, coverage needed, etc.).

Sequencers (analysts) receive the requests and prepare for the analysis, which includes:

- Searching the analytical laboratories' primer inventory for candidate primers.
- Designing and ordering primers (oligos) not in the inventory.
- Setting up, scheduling, and assigning DNA sequencing reactions to plates for either capillary electrophoresis (CE) or slab gel machines.

Sequencers use the primers in the laboratory, preparing and running the reactions, loading the completed reactions for analysis and setting up and running the analysis machines. The final output of the machines is a set of chromatograms, one for each reaction in the run. Sequencers edit and combine the chromatograms' sequences to construct consensus sequences of the clones submitted in the requests and provide the results to the requestors.

Management tracks the analytical effort expended on individual research projects based on the number of reactions performed within a designated time period. By comparing different time periods, a measure of operational and individual productivity is provided. The following DNA Sequencing Database system components support the analytical operations:

- Requesting: from scientist to sequencer.
- Integrated primer inventory and oligo design.
- Automated primer order generation and completion.
- Tube and plate format runs.
- Reporting results.
- Effort reporting based on history of completed reactions.

It is beyond the scope of this article to detail the many features and functions of DNA Sequencing Database. Instead, the important functions are highlighted, assuming that the reader has sufficient familiarity with computer database applications (as well as the DNA sequencing process) to recognize that an assortment of buttons, drop-down lists, and dialogs are

Chapter 11. A Flexible LIMS for DNA Sequencing Analysis   145

provided to assist the data entry and user communication needed to perform the requisite tasks.

Requesting: From Scientist to Sequencer

The DNA Sequencing Database is equipped with a password protection system that can differentiate the type of user, for example, scientist or sequencer. The graphical user interface (GUI) is tailored to the type of user, presenting them with specialized input forms. To minimize data entry, users can set preferences (research project, group, e-mail address, telephone number, etc.) that are used as defaults on the Request Entry form. Figure 11.1 provides an example of the Request form for new requests, showing the fields associated with the Request as well as a list of the clones (DNA) requiring sequence analysis.

Depending on the type of submission, scientists can select from three formats: Tube, Plate by Column, and Plate by Row.

The Tube format is unconstrained in that the number of clones is only limited by physical constraints of the system (e.g., RAM, disk space). The

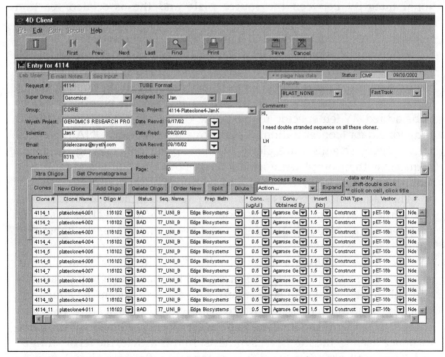

Figure 11.1. New Request input form. A requestor can edit clone names, concentration, and comments fields (comments field is visible after scrolling). All other fields are selected using drop-down lists.

correlation of submitted clone samples to samples prepared for analysis (synthesis reactions) is left to the discretion of the analyst. Although originally reserved for gel runs, Tube formatted reactions also can be assigned to plates and run using capillary electrophoresis (CE runs). The Tube format is principally used for low-throughput analysis, where any number of reactions can be performed for each clone. In this case, scheduling for analysis is automated, but analysis plate assignments are performed manually, using the drag-and-drop function in the DNA Sequencing Database's Run module. Moreover, reactions associated with clones from several requests (prepared by different analysts) can populate the same plate.

Plate by Row and Plate by Column formats designate that submitted clones are arranged by row or by column, respectively. When the source plate is full (96 clones), analysts are required to maintain the positional integrity of the clones on the analysis plates. In other words, each location on the analysis plate represents one reaction for the clone in the same location on the source plate. These formats are normally used for CE runs. The Plate by Column and Plate by Row arrangements provide for high throughput analysis (e.g., 96 locations filled) by maintaining the positional integrity between the source plate and the associated reaction (analysis) plates used to perform the analysis. Assignment of reactions to analysis plates is automated and accomplished by the system.

Wherever possible, drop down and popup lists are provided for the Request fields and the Clone columns' cells, to assist data entry and to ensure data integrity. Adding clones is facilitated by using existing clones as a template for the others and by simply entering the number of clones to be added. When specific request data (project, group, date required) and clone data (concentration, vector, coverage, species, DNA type, etc.) are required, scientists are alerted when saving and closing the form. Requests with incomplete information have pending status and cannot be sent for analysis until all of the required information is supplied.

The Request form consists of several pages, designated by folder tabs. One of these, Sequence input, provides the vehicle for scientists to submit a reference sequence to better define regions of interest for the submitted clone.

Scientists submit requests by selecting "Send Requests" from a menu, at which time a request number is assigned by the system to indicate it is available to the analytical staff (referred to as "sequencers"). Sequencers open the request and add rows to each clone, representing reactions needed to perform the analysis. The reactions are defined by the oligo sequence used to initiate DNA synthesis. In this article, the terms "oligo" and "reaction" are often used interchangeably. Because oligos serve to prime the DNA synthesis reactions, they also are referred to as "primers." Each oligo is assigned a unique integer number by the system (referred

Chapter 11. A Flexible LIMS for DNA Sequencing Analysis

to herein as "compound number") that is used to reference the material in DNA Sequencing Database's Primer Inventory and is used when ordering additional material from the oligo suppliers.

The Sequencer's Request form is similar to that shown in Figure 11.1, except that instead of listing clones associated with the request, it lists the reactions belonging to each clone in the request. The list contains the attributes (columns) of the reaction's clone as well as additional columns (oligo compound number and sequence name) to identify the reaction's oligo. This form has functionality to aid in generating rows of oligos and copy and paste functionality for quickly filling rows with data based on using existing oligos as templates. It also provides functions for ordering oligos, for designing primers, and for searching the Primer Inventory for candidates based on using a reference sequence or a clone fragment determined during analysis. These features are described in more detail in the next section.

Ordering oligos from the Oligo Synthesis Group is triggered manually from the Request form, after which the process is handled by synchronized system communication between DNA Sequencing Database and Oligos Database (another 4D Server-Client application for ordering oligos). Orders are automatically created in the Oligos Database and subsequently scheduled (when completed) for analysis by DNA Sequencing Database.

Integrated Primer Inventory and Oligo Design

Before January 2002, the Primer Inventory and the Design of Oligos were performed outside of DNA Sequencing Database by manual operations using third party software and other databases. Incorporating these functions into DNA Sequencing Database and integrating them into the Request process achieved large productivity improvements. In our experience, sample throughput has doubled and using this integrated approach has cut preparation time in half.

For example, because the Primer Inventory table in the DNA Sequencing Database contains all the primer material used in prior analyses (including their sequences), it is straightforward to provide the search functionality that finds primers whose sequences exactly and uniquely match reference sequence and/or clone fragment regions. (Note: *match* means that the primer sequence is the reverse complement of a portion of the reference sequence or fragment within the region of interest.)

In addition, adding reactions using the selected primers to clones comprising requests and scheduling them for analysis is literally a mouse click away. Performing these functions within DNA Sequencing Database is accomplished in about one or two minutes compared to hours for the nonintegrated approach. Along with these productivity improvements, the integration also has reduced costs by making more efficient use of the inventory, reduced the number of current oligo orders by as much as 80%, and provided for continuing reductions in the future.

To further support the analytical effort, DNA Sequencing Database supplements primer selection by integrating a Primer Design module in consort with the Primer Inventory module. This module provides oligo sequences for those regions in the reference sequence or clone fragment for which no matches in the Primer Inventory were found. It not only uses the unique sequence criterion to match the sequence in regions of interest, but also imposes constraints on the oligo sequence itself (GC % range; Tm range, stability, hairpin separation, base runs, matches at 3' end, adjacent homologous bases and repeats) to promote successful synthesis reactions.

Oligo selection for 30 regions in an 8.5 kbp reference sequence is typically accomplished in less than two minutes. Moreover, adding the candidate oligos to a request, ordering them in the oligos database and scheduling them for analysis is handled programmatically, again literally a mouse click away.

The Oligo Designer module can be accessed as a stand-alone function for application across requests or from within a Request. Both modes provide similar functionality. The former, however, supports adding oligos to several requests that contain clones that have the same reference sequence. Figures 11.2, 11.3, and 11.4 illustrate the three pages of the Oligo Designer form. Figure 11.2, the Sequence page, provides for the entry of sequence information. Here, the scientist delineates the region(s) of interest and provides the reference sequence. Alternatively, the sequencer can also use this page to provide a sequence fragment.

Figure 11.3, the Primer Search page, provides for the search for candidate primers, based on entering the minimum primer length. It also contains a mechanism for selecting and adding them to the clones in the request. Finally, it lists the criteria to be used for designing oligos in case primers cannot be found for some regions of interest. It provides editable regions of interest as well, defaulted to 500 base pair increments along the sequence. These criteria are used by the system in designing oligos (Figure 11.4).

The Oligo Design and Order page (Figure 11.4) provides another view and selection mechanism for adding and ordering oligos based on the regions of interest. In this view, each region is a row that contains drop-down lists of oligos candidates found in the Primer Inventory. No Match regions (no candidate primers found) can be selected for ordering, which invokes the Oligo Designer to determine the sequences of oligos to order based on the criteria entered on the Primer Search page. Note that two lists of oligos are shown, one for the forward direction (sequence) and one for the reverse direction (reverse complement of the sequence). Action

Chapter 11. A Flexible LIMS for DNA Sequencing Analysis   149

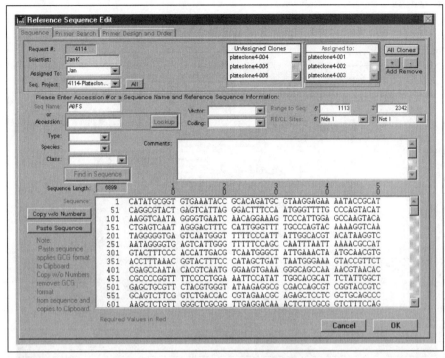

Figure 11.2. Sequence input page of Oligo Designer. The same reference sequence can be assigned to many clones; alternatively, the same clone can be assigned to many reference sequences. For unambiguous assignment and orientation of the reference sequence, some information is required as indicated.

cells are also color coded to differentiate regions with No Matches from those that have Add or Order oligo candidates.

The tabs on the form provide access to the pages. All pages provide a common entry area for request information as well as for selection of the clones in the request that pertain to the Add and Order oligo candidates.

Automated Primer Order Generation and Completion

Figure 11.4 shows that sequencers have the option of adding Primer Inventory oligos to regions of interest one by one or by treating them en masse. Sequencers can, if they choose, replace those selected using the drop down list of candidates. Regions of interest can be skipped or ordered as well by selecting "Skip" or "Order" from the Action popup.

Similarly, users can enter a forward direction prefix and a reverse direction prefix for their sequence names (all oligos are assigned unique

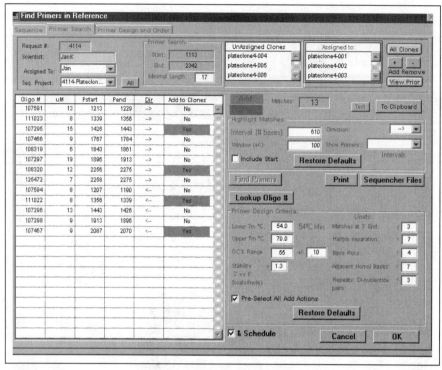

Figure 11.3. Primer Search page of the Oligo Designer. Positions and orientations of primers are displayed after primer matching (using the Find Primer button). Highlighted primers can be automatically added to selected clones. Default primer design criteria are shown on the right.

sequence names in DNA Sequencing Database) and Order All oligos for the No Match regions. Alternatively, sequencers can select regions individually by changing the action from "Skip" to "Order" and by then entering a unique sequence name.

Clicking the "Action" button results in the determination of the oligo sequences to be ordered and sets up the oligos to be added to the request, performing a variety of tests to make sure that they are appropriate for the request's clones. Closing the dialog by clicking "OK" adds the primers and ordered oligos to the clones in the Request, schedules them (optional), and generates orders in the Oligos Database (described earlier).

Completing orders is an automated process performed by DNA Sequencing Database in which the Oligos Database is checked periodically (30 min intervals). Completed orders result in updating DNA Sequencing Database and automatically scheduling the ordered oligos for analysis.

Chapter 11. A Flexible LIMS for DNA Sequencing Analysis   151

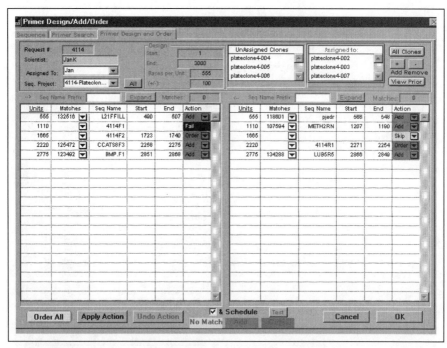

Figure 11.4. Primer Design and Order page of the Oligo Designer. If needed, the sequencer can change the default criteria (shown in Figure 11.3) to design some or all primers and add them to some or all clones.

In summary, integrating the Primer Inventory Search and Oligo Designer module with the Request and Order functions provides rapid, comprehensive support for the analytical function and workflow. Based on these features, sequencers are ready to prepare their reactions and schedule them for analysis.

Tube and Plate Format Runs

To provide for the laboratory preparation and running of reactions, the DNA Sequencing Database provides a Run module for assigning scheduled reactions to an analysis plate. This module, originally designed for handling gel runs, has been updated to handle capillary electrophoresis (CE) runs. CE has become the primary analytical vehicle in our laboratory.

In Figure 11.5, the Run input form provides a drag and drop interface for moving scheduled reactions (left list) to specific locations on the analysis plate (right list). For gel runs the locations are designated numerically (1 to 96), while for CE runs the locations are alphanumerically designated

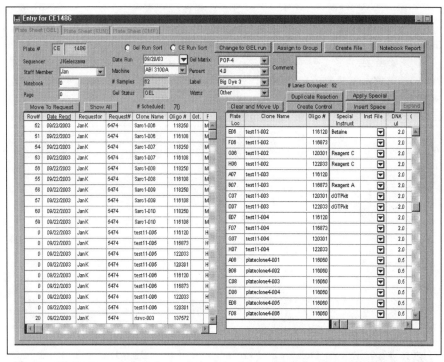

Figure 11.5. Run input form for assigning reactions to an analysis plate. Because many sequencers can share the same sequencing run, this form consists of a "holding buffer" (left-hand side of the form) where all reactions are temporary stored. Upon deciding the reactions to run, sequencer just highlights them and drags them to the right-hand side. When preparing the notebook report, the sequencer has an option to organize, in descending or ascending order, primers needed for easy retrieval from a storage system.

(A01 to H12). Provision is made for the analyst to add control samples, duplicate reactions, and designate special instructions for selected reactions.

The system also assists the analyst by calculating, based on the clone and the oligo concentrations, the required dilution factors, DNA volume, oligo volume, and TAQ mix for each reaction.

Once reactions have been assigned, the system outputs a Sample Sheet file that is compatible with the specific analysis machine selected on the form. Importing this file into the analysis machine sets up the machine for analysis. The machines supported by DNA Sequencing Database currently are ABI 3700, ABI 3100, ABI 373 and ABI 377, Other types of machines can be added easily to this list.

DNA Sequencing Database also provides the analyst with a hard copy Notebook Report of the assigned reactions list (right-hand list in Figure

11.5) for their research notebooks and for the preparation of the sequencing reactions at the bench. The Notebook Report contains all the calculated values needed to perform the setup of the DNA sequencing reactions. For high throughput analysis, DNA Sequencing Database automates this process, by creating the Run Plate record and assigning the reactions to the plate. Sequencers need only print their Notebook Reports and output the Sample Sheet file to proceed.

In general, runs can contain samples from several requests. In essence, the Run module provides a shared resource for several analysts to perform sequencing in parallel. The Run module input form also contains a page for completed runs. Functionality is provided on this page for duplicating failed reactions, modifying them with special conditions, adding them to the Request and scheduling them as reruns. Failed reactions are assigned failure codes that are used to assess causes of failure for further improving operations.

Reporting Results

The 4D DNA Sequencing Database also provides a Results module that displays results by Clone rather than by Request. Results consist of four components obtained from using Sequencher analysis software:

- The consensus sequence (or contig) derived from assembling the sequences obtained from the analysis of the reactions.
- The Overview—a location map of the specific reaction sequences used to build the consensus relative to the contig.
- The Sequencher project file containing the chromatogram reads, the analysts edits, and the details of the results.
- The protein sequences derived from the DNA consensus sequence results.

Sequencher is software for DNA Sequencing from Gene Codes Corporation. For detailed information on the features of this product, the reader is referred to the Web site (http://www.genecodes.com).

Reporting the results by clone supports the search for and grouping of duplicate or revised cloning results together as a basis for comparison. Scientists are primarily interested in the clones they submit for analysis rather than the Requests that contain them. This module provides requestors with pages displaying the consensus sequence and the overview. It also provides for downloading a copy of the Sequencher project file to the user's desktop and launches the Sequencher application for those scientists who want to review the analyst's edits and analytical results in depth.

Effort Reporting Based on History of Completed Reactions

Because it is a database application, DNA Sequencing Database stores information that is invaluable in assessing and refining the operations of the analytical laboratory. As a case in point, the use of the number of reactions run within a designated time period provides a measure of the DNA sequencing effort expended for Wyeth research projects.

This module uses the number of completed reactions per sequencer as the unit of measure of effort. Effort consists of three components:

1. Lab: fraction of effort to setup sequencing reactions.
2. Analysis: fraction of effort to verify and analyze the data.
3. Support: fraction of effort for shared functions (ordering, administration, maintenance, etc.).

The Lab effort is augmented by the time spent to perform the analysis (Analysis). Support is the time spent to perform administrative and maintenance functions shared across the DNA Sequencing Group projects.

Initially, each project is assigned a percentage of effort: Lab:Analysis: Support = 50%:50%:0%.

Figure 11.6 illustrates the Project Effort form used to view and report effort during a specified time period. The effort can be viewed for an individual sequencer as well as the entire group of sequencers (ALL) and for an individual project or the list of projects requesting analysis (ALL).

Analysts enter/edit the Lab and Analysis percentages for each project based on their actual level of time spent in each activity. For example, if the time to perform the analysis is twice the time to setup reactions for a particular project's requests and 10% of the analyst's time is shared across requests (Support), then Lab:Analysis:Support = 30%:60%:10%. The system calculates the support component for the analyst based on the Lab and Analysis entries (100% − Lab% − Analysis%). In this way, sequencers can differentiate difficult analyses from routine ones and also include the component for support.

Basing the effort on a measure of actual output of the department (number of reactions per project/total number of reactions for the specified time period), as opposed to only the time individuals charge to projects, provides an improved measure of productive effort. Management reviews this information to assess and control the allocation of resources to specific projects as well as to obtain a more objective view of individual performance within the group. Because the Effort Reporting is based on actual output, it can be utilized for accounting purposes as well.

Chapter 11. A Flexible LIMS for DNA Sequencing Analysis

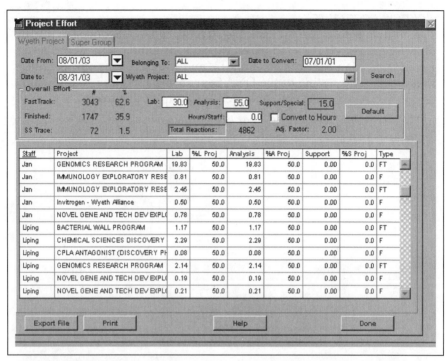

Figure 11.6. Project Effort report form. The effort expended on a specific project by an individual or whole group is based on the type of DNA sequencing service performed (finished or data). In addition, the effort is calculated by functional units (Super Group) in the organization.

Summary

The DNA Sequencing Database serves as an indispensable tool of the DNA Sequencing Group. It provides the means for capturing information with a minimum of manual data entry, and arguably performs the more important role of assisting analytical planning and design and automating time consuming manual operations, such as, ordering oligos and scheduling reactions. In addition, it provides management with a quantitative assessment of output as a useful component for improving operations.

DNA Sequencing Group personnel not only are enthusiastic proponents of the system but also continue to propose innovative features directed at improving their productivity and the quality of their work. The features described here are a result of this ongoing collaborative effort.

Acknowledgments

The authors would like to thank the staff of the DNA Sequencing Group of Wyeth Research at Cambridge (Mostafa Ait-Zahra, Katarzyna Bajson, Lora Haines, Jan Kieleczawa, Tony Li, and Liping Zhang) for countless suggestions and testing of the DNA Sequencing Database. Many thanks also to Charles Richard for continued support of this project.

12 Good Laboratory Practices, Good Manufacturing Procedures, and Quality Assurance in the DNA Sequencing Laboratory

Michele Godlevski and Thalia Taylor
GlaxoSmithKline, DNA Sequencing Facility, Research Triangle Park, NC

Introduction

From their inception, most molecular biology techniques and instruments were intended for use primarily in academic research labs. However, the DNA sequencing revolution from low throughput to high throughput capacity and the fruition of the Human Genome Project have enabled genetic technology to move deep into the diagnostic and pharmaceutical arenas. When technologies like this cross the boundary from research to medicine, they may become subject to the scrutiny of the U.S. Food and Drug Administration (FDA) and might then need to conform to good laboratory practices (GLP) or good manufacturing procedures (GMP).

Most laboratories, if not required, steer very clear of regulatory standards, adopting a "why do it unless you have to" philosophy. However, any core facility manager will admit that a large portion of their job involves trouble shooting and customer service. In addition, most DNA sequencing facilities experience a relatively high rate of employee turnover due to the repetitive nature of the work. Under these conditions, it doesn't take very long to come to the realization that troubleshooting is virtually impossible without taking measures to ensure some degree of consistency of laboratory practices. With the customer perpetually asking the question "are you sure someone in your lab didn't do something incorrectly?" or "are you sure a machine didn't fail?" the DNA sequencing facility manager is faced with the burden of proof that everything in the lab happened exactly as it should have. Moreover, if the DNA sequencing data eventually become linked to a labeling claim of a pharmaceutical, the FDA now gains the authority to walk into your lab and ask the same questions. If the FDA finds that your lab was not in compliance, they could assign a variety of penalties that ultimately could affect the profitability of your company, including disqualification of a study or a product recall. So, while it is not required that most DNA sequencing facilities conform to GLP standards, there is a benefit to be gained by utilizing GLP as a means of maintaining a quality in the overall DNA sequencing process.

Good Laboratory Practices and Good Manufacturing Practices

There is a common misconception in the nonregulated research laboratory that the FDA actually places scientific constraints on a GLP/GMP (collectively referred to here as GXP) compliant lab. In fact, the FDA simply asks that scientists make a written commitment to do what they say and say what they do. Their remit is not to limit innovation; it is to increase data integrity, product uniformity, consistency of outcome, and assuring the ultimate safety of the people who benefit from the technology. The FDA's role, simply put, is to protect the consumer from substandard or defective products by establishing minimum benchmarks of acceptability and reproducibility.

The genesis of government involvement in science and medicine dates back to the late 1800s, when "patent" medicines became popular. So-called "medicine men" traveled from town to town amazing crowds with their claims and selling their "cure-alls." These products, with such names as "Hamlin's Wizard Oil" and "Dr. Hammond's Nerve and Brain Pills," often contained harmful substances and even narcotics such as opium, cocaine, and heroin. Labels did not list the ingredients or warnings, leaving the public to make uneducated medical decisions for themselves. Product salesmen boasted of the ability of these substances to cure multiple types of aliments including serious conditions such as diabetes.

At the same time, the growing population in American cities led to an increase in industrial food production. This was a new and virtually unregulated industry. Use of toxic preservatives and colors as well as lack of sanitation practices provoked a chemist named Dr. Harvey Wiley to begin crusading against dangerous and poisonous food and drug products. Wiley's campaign and Upton Sinclair's novel, *The Jungle*, exposed the horrors of the Chicago meat packing industry and increased public awareness. Subsequent protests lead to the passing of the Wiley Act of 1906, also known as the Pure Food and Drug Act. The Wiley Act stated that drugs had to abide by standards of purity and quality set forth by the *United States Pharmacopoeia*, and prohibited the sale of adulterated and mislabeled food and drug products. The Act also was designed to enforce these regulations by creating the FDA and allowed punishment including seizure of products, fines, and even imprisonment.

Although the Wiley Act led to great improvements, there were still many flaws in the regulation of the food and drug industry. Most importantly, products could still be marketed without safety testing. One example of such a product was the "Elixir of Sulfanilamide." This elixir contained a solvent similar in chemical make-up to antifreeze and killed 107 Americans, including several children. The Sulfanilamide disaster dramatized the need to establish drug safety before marketing, so the Federal Food, Drug and Cosmetic Act was passed in 1938.

In 1951, the first prescription was dispensed. Before this date, patients could buy most drug products without a doctor's consent. Then, in the 1960s, Thalidomide, a new sleeping pill taken by pregnant women in Europe, resulted in thousands of babies being born with severe birth defects. Although Thalidomide was not approved in the United States, this event focused public attention on further strengthening U.S. federal regulations. The Thalidomide scare led to the Kefauver-Harris Amendments, which became effective in 1963 and stated that efficacy must be proven and adverse reactions must be reported and advertised. Another part of these amendments was the federal requirement and definition of GMPs (6). Before GMP regulations, a pharmaceutical could be contaminated by microorganisms or by other pharmaceuticals, could contain poisons, or could have contained varying amounts of nondisclosed inactive ingredients. With these regulations in place, the FDA was allowed to inspect manufacturing facilities and to force product recalls. There also were several revisions to the GMP regulations in later years that further refined these regulations to include medical devices.

Meanwhile, in a 1975 hearing before the Senate Judiciary Committee, Searle Laboratories was brought to court for allegations of improprieties in animal testing safety standards. The following year, subsequent hearings chaired by Senator Edward Kennedy resulted in the formation of a bioresearch monitoring program and the initiation of widespread inspections of nonclinical laboratories. The FDA found various forms of data integrity violations, including failure to have standard operating protocols and any system of quality assurance. In some laboratories, they uncovered falsification of data and exclusion of unfavorable test results. Out of these investigations came official GLP regulations in December of 1978. These regulations are applicable to any laboratory producing safety data in the support of products regulated by the FDA and include: food and color additives, animal food additives, human and animal pharmaceuticals, medical devices, biological and electronic products (6).

GLPs may apply to the DNA sequencing laboratory when the data being generated are utilized in any type of submission reviewed for safety by the FDA. The most challenging aspect of understanding the Code of Federal Regulations on GLPs is that they were originally written with animal research studies in mind. Interpreting these regulations as they would apply to DNA sequencing and genetic research requires some finesse, and access to regulatory expertise. GMPs apply to the DNA sequencing laboratory when DNA sequencing data are utilized to analyze a manufactured product, for instance, testing a lot of pharmaceuticals for the presence of microorganisms. In this sense, the DNA sequencing is a quality control measure, and it is bound by GMP standards. In any case, because the DNA sequencing laboratory produces a "product" (albeit an electronic product), applying many of the guidelines for GLPs and GMPs make trouble shooting and problem diagnosis much more efficient and effective. Thus, until there are Good Genetic Data Practices (GGDPs), the GLP and GMP regulations can be retrofitted for the DNA sequencing lab.

Standard Operating Procedures

Standard operating procedures (SOPs) are the written instructions that describe how GLP/GMP requirements are to be met. It is the expectation that the procedures in a GXP lab are performed in the same manner from person to person, from week to week. While human nature lures us to "tweak" and to pass on knowledge in an oral tradition, the FDA would like to know that every time a procedure is performed, it is performed in a consistent manner, and that any changes are validated, documented, and authorized. They also want assurance that every time a new person joins the laboratory, there is a reference document for them to refer to and be trained from. From the FDA's perspective, there must be a "letter of the law" in the laboratory. Failure to comply with the laws governing GLP/GMP regulations can result in imprisonment.

The guidelines for SOPs require that all of the following laboratory activities are described in an SOP. By requirement, SOPs should address one subject and be easy to understand, user-oriented, direct, and should clearly define responsibilities. The basic components of an SOP are (1) subject, that is, what the SOP addresses; (2) purpose, what the procedure is for; (3) scope, where the SOP applies; (4) references, what documents relate to this SOP; (5) responsibilities, who will be doing what; (6) definitions, terminology needed to understand the SOP; and (7) the procedure itself. SOPs are not lengthy documents, most are only one to four pages in length.

By content, a SOP should contain the following: (1) a descriptive title and purpose; (2) the date the SOP becomes operative and the edition

Chapter 12. Good Laboratory Practices

number; (3) the distribution of the SOP; (4) the person responsible for writing the SOP; and (5) the person authorizing the SOP. Other considerations for SOPs are that they should:

- Communicate effectively with short, active sentences and well-defined simple terms.
- Be sequentially directive (instructions listed in the correct order of performance).
- Be limited in the amount of background information.
- Be immediately available in the laboratory.

As they relate to the DNA sequencing laboratory, the Good Laboratory Practice requirements state that:

> A testing facility shall have standard operating procedures in writing setting forth nonclinical laboratory study methods that management is satisfied are adequate to insure the quality and integrity of the data generated in the course of a study. All deviations in a study from standard operating procedures shall be authorized by the study director and shall be documented in the raw data. Significant changes in established standard operating procedures shall be properly authorized in writing by management.
>
> Standard operating procedures shall be established for, but not limited to, the following:
>
> - Receipt, identification, storage, handling, mixing, and method of sampling of the test and control articles
> - Test system observations
> - Laboratory tests
> - Data handling, storage, and retrieval
> - Maintenance and calibration of equipment
>
> Each laboratory area shall have immediately available laboratory manuals and standard operating procedures relative to the laboratory procedures being performed. Published literature may be used as a supplement to standard operating procedures. A historical file of standard operating procedures, and all revisions thereof, including the dates of such revisions, shall be maintained (6).

In the DNA Sequencing laboratory, this means that:

- Samples must only be handled and processed by qualified personnel.
- All training records and employee qualifications must be kept on file and be readily available.
- Samples must be processed in a specific compliant location and stored in designated areas according to the stage of processing.

- A method of naming samples by unique identifiers must be established and utilized.
- Samples must be labeled with a unique identifier on the 96-well plate or sample tube.
- Conditions for thermal cycling and electrophoresis must be standardized.
- All machines involved in the processing of samples (including refrigerators, pipettes, thermal cyclers, and sequencers) must be calibrated and their performance must be verified on a regular basis.
- Procedures for data processing, distribution, archival and retrieval must be dictated and set up to minimize the possibility of data altercation.

Because SOPs are prescriptive rather than descriptive in nature, it is possible and advisable to keep them fairly brief. It should be obvious that issues like a sample naming discrepancy could cast doubt over data integrity. Any deviations from the SOP, regardless of consequence, must be fully documented and correlated to the affected data.

Quality Assurance

The terms *quality assurance* and *quality control* are often confused. Quality assurance (QA) is the internal monitoring program a company administrates to make sure that it is adhering to GXP practices. It is a requirement of GMPs to have a QA program in place (6). It is the responsibility of a quality assurance group to ensure that (1) SOPs are available, effective, and are being followed; (2) SOP edition numbers are correct and obsolete editions are withdrawn from circulation; and (3) SOPs are reviewed for unexpected interpretations and for missing elements. The remit of the QA department is to maintain records and protocols (document control) as well as to inspect the laboratory and make recommendations for problem resolution. A QA program should be designed to establish the quality of a product or output by identifying critical aspects of the process, and then monitor quality and manage changes to the process.

Quality control testing is considered to be a part of the entire QA program. Quality control measures address the evaluation of raw materials, packaging, labeling, and finished products to prevent the distribution of a defective or substandard end product (1). While the original language of the GMP regulations is directed at manufacturing, in the DNA sequencing laboratory, the interpretation of "defective or substandard end product" is certainly applicable. For example, consider the case of an instrument failure resulting in blank samples files—a blocked well or a bubble on a gel-based system—or a blocked capillary or an injection failure on a capillary based system. Then consider the type of client for which a blank sample might be a real result, those studying the effects of pharmaceuticals on a viral infection, for instance. It would be considered a "defective or substandard end product" to publish "blank" data files generated due to a blocked well or an injection failure. However, it is not acceptable to simply withhold that data or to decide to rerun it on a hunch. Regulations would require that you have a quality control diagnostic for demonstrating that the machine was not functioning properly and that the results were affected, so that the withdrawal of the data was not misinterpreted as suspect.

As a second example of the need for quality controls in DNA sequencing, consider the occurrence of samples affected by lane leakage or instrument-related background noise with regard to the laboratory looking for heterozygotes in single nucleotide polymorphism (SNP) detection. Publishing data to these clients without having quality controls in place to determine sample integrity would be producing a defective or substandard end product and misleading data. Even in the nonregulated lab, some simple quality control measures could verify that peaks underneath peaks were real data.

Some suggestions for quality control measures in the DNA sequencing lab are listed below. While this list is not all-inclusive, nor are all required, these measures support the production of accurate data, unaffected by machine-related failures.

Gel-Based System

- Polymerization control: Retain a small sample of acrylamide from each gel poured in a test tube to assure polymerization is complete for each gel poured.
- Lane marker: Alternating molecular weight lane marker 5th dye spiked into sample loading buffer assures proper lane tracking and also indicates lane leakage. Verifies that electrophoresis occurred in each lane, even if there is no sample present.
- Positive control samples (pGEM) in outer lanes: Especially in a core facility, this measure assures that samples of high purity with good primers were readable by the system, thereby ensuring that blank samples are real results in so far as the functionality of the system for that run is concerned. It also is appropriate to define the minimum requirements of average signal strength and Phred quality scores for those controls.

Capillary-Based System

- Capillary functionality control: Run a plate of positive controls (pGEMS) as the last plate of the day (or on a routine basis) to

verify that capillaries are not blocked and specify the minimum requirements of signal strength and Phred quality scores for those controls.
- Injection control: Low molecular weight marker spiked into loading buffer verifies that electrophoresis occurred in each lane, even if there is no sample present. Indicates sample carry-over/contamination issues.
- Contamination tests: On a periodic basis, run a plate of water samples alternating with positive controls (2 columns of controls, 2 columns of water, etc.). If there is carry-over on the needles or photon hop between samples, correlating patterns will emerge in the water samples.

One of the most important aspects of a quality assurance program is having a system of reporting and following up on aberrant data. If a machine fails, for instance, in the nonregulated lab, this communication may be done by word of mouth, e-mail, or—worse yet—by Post-it note. When the client hears of a machine malfunctioning and then asks, "Were my samples affected by this problem?" the facility manager not only should be able to answer the question but must be able to document it as well. On the other hand, if there is a formal and standardized system of documentation, there is no ambiguity or miscommunication. A standardized error reporting and resolution form gives the option of reporting things that require attention before the data is published. This "follow-up form" gives the option of reporting issues like samples with low signal strength, which are not the direct result of a quality issue in the DNA sequencing process *per se*, but may need follow-up, from a customer service perspective. The follow-up form also serves as documentation to report issues that are urgent in nature, like a machine failure or a purification error, which would require immediate follow-up and remedy of the problem within the lab. Serious issues are followed up also with a corrective action form, which documents what will be done to prevent the problem from reoccurring in the future. Any resultant change in an SOP or guidance document would then be managed by the document control department of quality assurance to control distribution.

Customer Service

If the QA system is effective, perhaps the naive observer would presume that overall laboratory process should be close to flawless. While that would be a nice argument for justifying the effort to put a quality control process into place, there are still bound to be problems that arise. The merit of having a quality control system is not that it certifies the system as flawless, but rather that it gives the laboratory the tools to diagnose and remedy many problems before they affect the customer.

By virtue of their complexity, highly automated processes like DNA sequencing are wrought with subtle problems that seem to present frequent challenges. At the same time, neither SOPs nor personnel are perfect, and occasionally, a flaw is found in the process or in the way the process is being interpreted. For this reason, a problem (or perceived problem) reaches the customer and an inquiry is made. Often this inquiry follows the familiar theme of "my reactions worked last time, why didn't the exact same reactions work this time?"

As a customer service professional (which is hidden in the fine print of every DNA sequencing core facility manager's job description), it is prudent to recognize that customer inquiries should be taken seriously and given due consideration. Moreover, all inquiries should be documented. To aid in this process and to ensure that trouble shooting is carried out in a methodical manner, a "customer service inquiry form" is a useful tool. On this form, all of the intake data are recorded surrounding a customer complaint. It is helpful to ask standardized questions (like, "what template preparation method did you use?" "how did you quantify your template?" and "how much template did you add?") to be sure to cover all of the basic questions with each customer. It is a good idea, of course, to attach supporting documentation, for example, trace files and array or gel images, to this form.

Often the results of this investigation require customer counseling. While the subject of customer service alone fills bookstore shelves, it is wise to remember to be responsive, empathetic, and to assure the customer that you will make an objective investigation into their claim. Obviously, this approach requires a humble, composed demeanor, as defensiveness will only arouse suspicion that there is something being concealed.

Corrective and Preventative Action

The old adage "the customer is always right" reflects an attitude that should be adopted in the field of customer service. Most facility managers concur that core facility troubleshooting most often results in changes that need to be made to template quantity or quality. However, occasionally, despite the best intentions of quality assurance, an investigation of a customer complaint reveals a weakness in the system, and corrective actions in the process are necessary.

Corrective action plans should have defined timelines of resolution and should specify the type of changes that would require re-validation. They should specify who is responsible for complaint resolution, what type of complaint justifies an investigation, and what actually defines a nonconformance. It also should be very clear in the corrective action documentation which corrective actions are temporary "quick fixes" and which are true permanent corrective and preventative actions. Any corrective action should be appropriately documented, so that it may be referenced in the future.

A corrective action form is useful for this purpose in the DNA sequencing laboratory, and all communications regarding the complaint also are documented on the customer service inquiry form. The corrective action form includes:

- The affected data and customer.
- The nature of the problem.
- The results of investigation done regarding the problem and all supporting data.
- Any corrective action necessary, including changes in SOPs.

In the DNA sequencing laboratory, a commonly heard complaint is that all of a particular person's samples came back as "blanks." If the investigative data collected show that there was no instrument-related or process-related reason for this failure (based on quality control data and data gathered from the client on sample details), it may be concluded that the samples were blank by virtue of sample integrity (template or primer quality, or incorrect vector being used, etc.). However, there are some occasions when the result of the investigation might be that there was reasonable cause to believe that the blank data was because of an instrument or process failure. In this case, a root cause analysis should be done. If the failure was the result of an assignable cause—a one-time event like an instrument malfunction—the only corrective action required is that service on the machine should be performed and the sample should be reprocessed. However, if, in your investigation, you determine that there is a larger problem present, such as a weakness in protocol that doesn't catch a regular occurrence (like injection failures), then corrective action and corresponding revalidation should be performed and documented. In the case of an identified process or instrument failure, problem resolution should be communicated to the complainant in writing and to any other possibly affected clients. After a corrective and preventative action is taken, follow-up meetings with staff members and process-improvement brainstorming sessions are recommended. Furthermore, it is considered good customer service practice to acknowledge the issue to all clients, the data it affected, and its resolution. It is helpful to keep a list of corrective actions resulting in SOP changes and their effective dates on an internal Web site or on an internal records management system. Posting this information alleviates the concern that *all* data were affected.

Chapter 12. Good Laboratory Practices

Many times the problem has a very specific signature, like T-blobs in the data at a specific location. Giving the customer the information about exactly what the problem was, what the symptoms were, and what remedy was taken and exactly when it was taken alleviates customer anxiety. Such information also prevents the inevitable telephone call from a client asking if the T-blob problem you had last week somehow prohibited someone's reactions from working at all for the last three months. Thankfully, the problem at hand is rarely that pervasive, but to the customer, it is the first perceivable possibility.

Facility Requirements

A common erroneous belief about GXP work is that it can just be done as a smaller "project" in a noncompliant lab. GXP regulations require dedicated space and equipment that are only used for GXP applications. There are basic safety regulations regarding lighting, ventilation, air filtration, plumbing, maintenance, and building access that most large industries subscribe to company-wide. However, there are more specific specifications that set the GXP lab apart. For instance, all pipettes and instrumentation in the GXP lab need to be calibrated and be on preventative maintenance programs. There should be dedicated refrigerators and freezers that are monitored for temperature stability over time. Data must be published in a secure environment that is safe from possible manipulation. When an FDA investigator inspects the lab, he or she asks about the history of the facility and what type of work is done there. Having noncompliant work done even in one section of the same lab would provide ample opportunity for questions about the integrity of the data being produced there.

GMP regulations state that:

> Any such building shall have adequate space for the orderly placement of equipment and materials to prevent mix-ups between different components... to prevent contamination. The flow of components... through the building or buildings shall be designed to prevent contamination.
>
> Operations shall be performed within specifically defined areas of adequate size. There shall be separate or defined areas for the firm's operations to prevent contamination or mix-ups as follows: Receipt, identification, storage, and withholding from use of components... pending the appropriate sampling, testing, or examination by the quality control unit before release for manufacturing or packaging (6).

While this requirement might seem prohibitive to the nonregulated lab interested in doing GXP work, sequencing contract houses routinely perform regulated and nonregulated work in the same lab. The difference is that the *nonregulated work is done to GXP standards, even it is not required*. Why would a contract sequencing firm or just the average core facility want to work to GXP standards if there was some work that did not require it? The rationale is that confidence in QA can be directly correlated to consumer (and investor) confidence. This is true to the extent that most contract houses now ascribe to meet International Organization for Standardization (ISO) 9000 standards (7), which is a program that certifies a corporation in accordance to the level of QA it provides.

Training and Personnel

Another compelling consideration for the nonregulated lab is the task of maintaining consistency and quality in a staff-training program in an environment with high staff turnover. GXP regulations require that management maintain a written record of staff qualifications, training, experience, and job descriptions for all personnel. Especially in a high throughput environment marked by time pressures and deadlines, the FDA wants to know that the person filling in for someone has already had the necessary training experience to perform the assigned function. This requirement also extends beyond the laboratory to the QA personnel and even to management. There have been cases of serious FDA violations in which the top-level management, not the individual workers involved, went to jail for failure of the lab to comply. Certainly, any person whose signature is required on documentation should be trained once a year in current GLP and GMP regulations. It is also important that all levels of management in a regulated laboratory be trained in GLP/GMP requirements so that the policies they make do not conflict with the standards that must be adhered to. Training protocols and proficiency checklists should be designed and utilized during the training of all new employees. While it may seem resource-intensive to devote one month of one person's full-time responsibilities to training a new employee, that dedication can ensure that the employee who emerges is fully trained and competent.

Reagents and Solutions

Without a doubt, reagents and solutions are probably the most likely starting points in a typical trouble-shooting investigation in a DNA sequencing facility. Cause and effect are the teachers that groom most laboratory scientists to become much more conscientious about aliquotting from stock solutions and recording batch numbers. The GLP requirements are relatively straightforward in this regard:

All reagents and solutions in the laboratory areas shall be labeled to indicate identity, titer or concentration, storage requirements, and expiration date. Deteriorated or outdated reagents and solutions shall not be used.

Upon receipt and before acceptance, each container or grouping of containers of components, drug product containers, and closures shall be examined visually for appropriate labeling as to contents, container damage or broken seals, and contamination (6).

Maintenance and Calibration

GLP regulations state that:

Equipment shall be adequately inspected, cleaned, and maintained. Equipment used for the generation, measurement, or assessment of data shall be adequately tested, calibrated and/or standardized. The written standard operating procedures required under Sec. 58.81(b)(11) [Page 309] shall set forth in sufficient detail the methods, materials, and schedules to be used in the routine inspection, cleaning, maintenance, testing, calibration, and/or standardization of equipment, and shall specify, when appropriate, remedial action to be taken in the event of failure or malfunction of equipment. The written standard operating procedures shall designate the person responsible for the performance of each operation. Written records shall be maintained of all inspection, maintenance, testing, calibrating and/or standardizing operations. These records, containing the date of the operation, shall describe whether the maintenance operations were routine and followed the written standard operating procedures. Written records shall be kept of nonroutine repairs performed on equipment as a result of failure and malfunction. Such records shall document the nature of the defect, how and when the defect was discovered, and any remedial action taken in response to the defect (6).

Rules for maintenance and calibration are probably the most straightforward part of the regulations to implement and to justify for the nonregulated lab. Even in the purest of academic environments, it is recognized that a miscalibrated or poorly maintained instrument wastes time, effort, and money. Most instrumentation companies offer service and preventative maintenance contracts, and most organizations even have in-house instrument repair and calibration employees for this reason.

Equipment Qualifications

While the disclaimer on the model 3700 sequencer at this time still reads "for research use only," it is clear by recent actions that Applied Biosystems has acknowledged and addressed that DNA Sequencers are being used to produce data for GXP environments. In early 2002, Applied Biosystems announced equipment qualifications support on the 3100 DNA Sequencer.

Equipment qualifications include installation qualification (IQ), operational qualification (OQ), and performance qualification (PQ). The purpose of these qualifications, respectively, is to document that (1) the machine is installed according to manufacturer's specifications and complies with all local safety, fire, and plumbing codes, (2) the machine is operating as it is supposed to and is free from any mechanical design defects, and (3) the reliable and reproducible performance of the equipment under both normal and less than optimal case conditions can be documented. IQ/OQ/PQ equipment qualifications define what to verify and measure, how to verify and measure it, how many samples are required, when the samples are to be taken, and what the parameters are of the desired output.

Process and System Validation

The term *process validation* in manufacturing is defined as demonstrating that the process does what it is supposed to do from the process development phase through the production phase, and continuing on to repeated batches of product. Process validation can be considered a QA tool also, because it establishes and verifies the standards of quality that are to be maintained (6). Process validation includes the equipment qualifications as described above. "System validation" usually refers to demonstrating the same for the computer systems and software that support a process.

In general, processes and systems must be validated so that there is a high degree of assurance that the process or system consistently meets predetermined specifications. System or process quality cannot be assured through the finished product inspection or testing; it should be verified through validation. Validation demonstrates that there is verifiable output, and that the risks versus benefits have been addressed. Validation serves to ensure customer satisfaction, cost reduction, improved product quality, and compliance to regulatory requirements. Validation demonstrates that the process or system works in a given environment, with trained operators, and standardized approved specifications and instructions. Any exceptions in the process or system or expected failure rates and conditions should always be addressed specifically. They should not simply be omitted.

The first element of system validation for a DNA sequencing facility

Chapter 12. Good Laboratory Practices

is a system flow chart of the overall process. The basic components of the flowchart are:

- System input
- Hardware/software used
- System output
- Referenced SOPs

The system flowchart is a part of the system description, which includes the model numbers of all instruments and computers, and the version numbers of all operating systems and supporting software. This document also describes each component's functionality and gives details about data storage locations and permissions.

The validation plan describes the validation process that demonstrates the system does what it is supposed to do and produces the electronic data as expected. It describes the deliverable items that should be produced during validation without error. Also listed in the validation plan are the validation team members, any referenced SOPs, methods, or guidance documents, and the revision history for the entire system. The actual validation protocol describes the scope of the validation, the individuals responsible, expected results, and the acceptance criteria of those results. Also included in the protocol are a description of an error resolution process and a statement of assumptions, exclusions, and limitations.

When the validation protocol is executed, the results of each validation step are analyzed and either approved or rejected. Rejected results undergo a failure analysis and corrective action before the validation process is restarted. Validation provides assurance that predictability, consistence, and conformance are all maximized.

In addition to validation testing, a common practice in the DNA sequencing laboratory—although not formally documented as such—are "ruggedness testing" and "robustness testing." Ruggedness testing is the validation procedure that compares the exact same procedure on different machines, with different operators. The purpose of ruggedness testing is to define a degree of acceptable variability between different machines and between different operators before data would be rejected. For instance, when a new dye terminator removal product is tested for use in a lab, it would be prudent to determine that all machines would give a certain minimal signal strength and Phred of more than 20 bases, but it would also be important to know that the variance between different machines and different operators was not greater than a certain percentage. Robustness testing is defining the limits of sensitivity of an instrument or reagent. An example of robustness testing would be optimizing the minimal dilution of sequencing enzyme that still produces a benchmark signal strength and Phred score.

Change Control

Fortunately, our field of expertise is flooded with technologies that tirelessly aim to sculpt greater efficiency into the DNA sequencing process. These new technologies often prompt validation testing and a resultant change in SOP. In addition, most software released today does not stay a viable product for long at version 1.0. Software developers have the unenviable duty of incorporating long wish lists into functional software improvements. While most software upgrades usually correlate to a process improvement, they also require documentation and tracking in a regulated environment.

Upon initiation of a brand new GXP system, a configuration management plan is established to describe the original system of all the computer hardware and software needed to process samples from start to finish. This configuration is referred to as the "system." The configuration management plan is used also to designate a descriptive plan of how changes will be reflected in the version number of the system. Each subsequent change then requires a change control document, detailing the change. Hardware upgrades and/or additions can be noted by an increase in the numeral in the first position. Software upgrades and/or additions are noted by an increase in the numeral in the second position. Other changes such as revisions in the templates, interfacing applications, or storage areas are noted by an increase in the numeral in the third position. Thus, if the system 2.0.0 (consisting of an ABI 373-XL DNA Sequencer and a Power Macintosh 7200/120 computer, operating on Mac OS 8.1, with Version 2.0 of Data Collection, and Version 3.0 of Sequencing Analysis), changes to an ABI 377-XL DNA Sequencer, the system version would change to version 3.0.0. If the Data Collection software changes to version 2.6, the system version would become 3.1.0. If the shared area where the data are stored migrates to a new server, the system version would now become 3.1.1. All of these changes would require change control documents and revalidation of the system.

Data Collection and Archiving

One of the most difficult aspects of validating a gel-based sequencing system is the concept of lane tracking lines. The fact that tracking lines can be moved and the resultant data changed causes quite a bit of extra documentation to demonstrate that the data, as published, remained unchanged and unaltered. Unfortunately, in gel-based sequencing software, the mere act of opening a gel file changes the modification date of the file. This critical element reflects the fact that the technology was designed with a "research only" intent; most software designed with a

Chapter 12. Good Laboratory Practices

regulatory application focus has a tighter reign on document change control, and opening a file without making changes does not alter the file date. The newest sequencer systems at this time of this publication not only have the integrity of one-sample-per-capillary but also have audit trail log files. Nonetheless, the gel-based system can be validated, but it requires a strict and thorough paper trail.

To document that the data are collected and are not altered, trace files can be printed out and the gel files can be stored in a secure location. Password protection also can be added to access the computers from which one could change the data. Laboratory information management systems (LIMS) often make system access control much easier, and many provide audit trails of all records, which is extremely helpful in the regulated environment. However, LIMS systems must be off-the-shelf products or they must be officially validated by their developers. An off-the-shelf LIMS system must still be validated within the whole computer hardware/software system.

Documentation and Electronic Records

The DNA sequencing process resembles most quality control laboratories in that wet chemistry samples are received and electronic data are produced. On the wet chemistry end, regulations require that all data must be recorded in a formal laboratory notebook as they occur. Laboratory notebooks should be prenumbered, should have prenumbered pages, and should be bound. Notebooks should be assigned to individuals by task to be performed, especially for repetitive tasks. Logbooks for reagents should be used to track lot numbers and usage of reagents. Logbooks also should be prenumbered, should have prenumbered pages, and should be bound or not removable.

Any calculations that are done should be documented in the laboratory notebook, and should include units of measure, conversion factors, and each step of calculation. Rounding of numbers is allowed for end results, but rounding procedures should be defined in SOPs. Electronic spreadsheets that perform calculations are allowed as long as they are validated to be functional and are locked.

All test data, validation data, and raw data should be included in the laboratory notebook. Required documentation includes the unique sample identification information, the date the samples were received, the date the samples were processed, and the date the results were posted.

In the traditional laboratory notebook, data should be written in black permanent ink and incorrect observations should be crossed out with a single line. All entries should be signed by the person writing the notebook and then reviewed and approved by another person who witnessed the data. There should be a signature file available to identify and verify all signatures. All entries also should be dated in a specific and unambiguous format, for instance, 21-Jul-2002. Unused portions of the page should be crossed out with a diagonal line. Changes in conclusions should be reviewed and approved in writing.

Unacceptable practices include (1) transcribing raw data from one source to another, including recording results on scrap paper first and then copying the data into the notebook; (2) obliterating any data, including erasing or using white-out; (3) back-dating; and (4) using anything but a permanent blue or black ink pen to write entries.

The regulation says that electronic files must be trustworthy, reliable, and generally equivalent to paper records (6). Appropriate references to all electronic files must be included in the laboratory notebook. For more details on LIMS systems, please refer to the Code of Federal Regulations 21, part 11 (available at http://www.fda.gov).

Inspections and Audits

Most organizations having any dealings with the FDA have their own QA unit. As a service and job responsibility to the company, QA professionals offer in-house audits. Labs should be taken advantage of this service on a regular basis. The audit team should be independent and objective, providing feedback that can be utilized to make sure a laboratory is compliant. Outside auditing companies can be hired to perform this function also. If nothing else, being audited assures that the necessary documentation is kept up to date and is readily available. As the saying goes in the regulatory arena, "if it wasn't written down, it didn't happen." Establishing a regular schedule of internal auditing ensures that all documentation is ready for an FDA inspection, should one ever occur. Internal audits should be conducted on a yearly basis at a minimum. When an audit occurs, it should be specifically defined to a particular area and should include a written schedule with a checklist of things to be verified. After the audit, there should be a debriefing and a report filed with suggested actions for improvements. After the improvements are made, a final report is filed that should be read and approved by management.

If the FDA comes in for an inspection, they will be checking for some of the following areas of noncompliance: (1) inadequate procedures that might not be fully descriptive or fail to define specific steps or responsibilities; (2) missing procedures that are being performed but the corresponding document cannot be produced; (3) nonconformity between what is done and what is described in the procedure; (4) operating to unapproved or obsolete SOPs; (5) missing, incomplete, or inconsistent records; and (6) aberrant or atypical results. The FDA may not see your internal audit reports but can look at previous FDA audits of your lab.

When the FDA comes to inspect your lab, they will use a standardized Quality System Inspection Technique. They are supposed to visit each company biannually and may visit without warning that an inspection will take place. In preparation for an audit, you and your staff should become familiar with the Code of Federal Regulations (specifically GLPs and GMPs) that apply to your laboratory. You should be sure that all procedures and documentation are readily available for review.

If the FDA finds areas of noncompliance, they can issue a regulatory letter, in which they site issues that should be addressed but are not severe in nature. The FDA also can issue a "483" letter of noncompliance that lists official actions indicated for each item that must be addressed within a specified time frame. If the issues are not sufficiently addressed, a warning letter may be issued that cites the legal action that will be taken by the FDA if the items of noncompliance are not remedied. Often, failure to reply to a warning letter results in actions that will have an impact on the entire company, such as significant fines, product recalls, or shutting down certain parts of the business.

Costs and Other Considerations for Implementation

In any industry, the financial cost of having a product recalled is enough to argue the necessity of QA standards. Lost time and effort can be easily justified as a worthwhile rationale. However, the loss of consumer confidence is a factor not to be overlooked. Even if the company rebounds, the damage from the negative publicity is sometimes difficult to overcome.

From the microcosm of the nonregulated core services facility perspective, there are certainly parallel effects and justifications. Putting quality controls in the process ensures that problems are corrected before they cause data recalls, lost time, reagents, and work. Having a systemized approach to documentation may sound like a lot of extra paperwork, but when you are faced with a serious trouble-shooting problem, you may find that it was worth the effort. If nothing else, if you take a look at the philosophy of the way a DNA sequencing laboratory needs to be run to suit the FDA, there may be some best practices that would justify a little extra time, effort, or cost.

Information Resources

- Quality Assurance Group of the United Kingdom
- Society of Quality Assurance in the United States

- International Society of Quality Assurance
- http://www.fda.gov

Acknowledgements

I would like to thank the following members of GlaxoSmithKline corporation: Tish Brown, Scott Kimbrough, Ken Gray, Scott Brown, Helen McNulty, Donald Garbarz, Daniel Burns, Michael Weiner, Nigel Spurr, Ganesh Sathe, and Ian Delmar for their support and critical review of this manuscript.

References

1. Avis, K.E., Wagner, C.M., and Wu, V.L. 1999. *Biotechnology: Quality Assurance and Validation*. Interpharm Press, Inc. Buffalo Grove, IL.
2. Carson, P., and Dent, N. 1990. *Good Laboratory and Clinical Practices: Techniques for the Quality Assurance Professional*. Butterworth-Heinemann Linacre House. Jordon Hill, Oxford, UK.
3. DeSain, C., and Sutton, C.V. 1996. *Documentation Practices: A Complete Guide to Document Development and Management for GMP and ISO 9000 Compliant Industries*. ADVANSTAR Communications, Inc. Duluth, Minnesota.
4. Levin, M.T. 1996. *How to Fail an FDA Quality Audit: A Look at Some of the Causes of Failure to Comply With FDA Quality Regulations*. Mort Levine, Inc. Natick, MA.
5. United States Food and Drug Administration. April 1996. Code of Federal Regulations, Title 21, Part 211. *Current Good Manufacturing Practice For Finished Pharmaceuticals*.
6. United States Food and Drug Administration. June 2001. Code of Federal Regulations, Title 21, Part 11. *Electronic Records; Electronic Signatures*.
7. International Organization for Standardization ISO 9000 documentation. Available at: www.iso.ch/iso/en/CatalogueListPage.CatalogueList. Accessed February 2004.

13 Future of DNA Sequencing: Towards an Affordable Genome

Kevin McKernan
Agencourt Biosciences, Beverly, MA

Introduction

With the current state of technology, the cost of one whole genome shotgun read is approximately 75 cents, and the estimated price tag for sequencing of the entire human genome is in the range of 20 to 40 million dollars. The expected evolutionary improvements on existing capillary array electrophoresis (CAE) technologies may bring these numbers below 10 million dollars. It is doubtful that much could be done to pull these numbers below the $1 million mark. For sequencing to move into the clinical market and for it to replace genotyping applications, the price needs to fall below $100,000 and preferably to about $10,000. Currently, there is active discussion targeting the $1000 genome and, even though these ambitious goals have been rallying for the technology industry, one must keep in mind what cost is needed for a disruptive change in the industry. Currently, 6 Gb of RAM (the amount of RAM preferred for fast alignment of re-sequencing reads to the human genome) costs more than $1000, so driving the cost of a human genome below this in the next three years may be incongruent with the rate of improvement in computation projected with Moore's law. RAM is of course reusable and the computational problems are addressable with current computer clusters, but the current computer farm is not the bench-top solution envisioned in the clinical implementation of the $1000 genome.

Less riskier technological advances are being proposed that may enable $100,000 to $1,000,000 human genomes in the next three to five years. Sequencing costs in this price range would enable genotyping, single nucleotide polymorphism (SNP) discovery, strain identification, karyotyping, and expression profiling to be cost effective using sequencing technology.

Incremental Improvements of Existing Technologies

The current leading technology for high throughput DNA sequencing is the Applied Biosystems (ABI, Foster City, CA) 3730xl 96-capillary automated sequencer. Under optimal conditions it is capable of generating of up to 2.5 Mb/day of $Q \geq 20$ quality bases (10). Based on ABI's evolutionary approach, it is possible that within the next three years a 3770 will be available (370, 373, 377, 3700, 3730-progression of ABI instruments [see also Chapter 9]) with an expected 1.5- to 3-fold increase in throughput. Polymers that enable faster separation times (24) and 384-well capillary arrays are likely to be installed in this time frame. The impact on the price per read would be substantial if the instrument price remains constant as with current sequencing technology; the price of the amortized cost of the sequencer is close to 50% of the price per read. The reagents usually represent 20% of the cost per read. As a result, the advertised reagent savings for each new platform should be evaluated with caution because ABI has a history of selling instruments based on this promise, only to raise the reagent prices few months after installation of the instrument.

Even if the sequencer is given away for free, and other materials continued to be stretched, the cost would still not fall much below 20 cents per read or an $8 million per human genome. The inherent fixed cost component of CEA sequencing is the overhead associated with handling 20 million clones and 40 million reactions with maximal plate densities of 384 wells. Candidate $1000 genome technologies promise to reduce this by utilizing an alternate library construction methodology or bypass it altogether.

After even a cursory review of current sequencing technology costs, it becomes apparent that most incremental improvements, although inventive and promising, will have marginal impact on the price of sequencing a human genome.

Sequencing Reaction Miniaturization and Capillary Thermal Cycling

Several proposals have been funded to pursue microscaling of thermal cycling and sequencing reactions, and one such integrated instrument is now commercially available (Parallab 350, Brooks-PRI Automation; http://www.brooks.com/pages/288_parallab.cfm). It appears that the driving principle behind building the Parallab instrument is the emphasis on the integration and reduction in reagents' costs through microscaling. Microscaling and even nanoscaling (microliter to nanoliter reactions)

Chapter 13. Future of DNA Sequencing

the dye-terminator reagent can save tremendous amounts of money, assuming that Applied Biosystems, with its dominant intellectual property position, doesn't simply raise its price structure. With the 2004 list price of approximately $2.00 per genome center reaction and $8.00 per list price reaction, most genome centers have engineered dilution buffers that allow 50- to 100-fold dilution of the dye-terminator reagent. As a result, the dye-terminator mix is no longer the cost driver in high-throughput (HT) DNA sequencing pipelines.

Thus, technologies aimed at delivering a 50-fold BigDye™ dye-terminator dilution utilizing capillary cycling technology are unfortunately a bit late. The much advertised benefits of faster cycling times are unfortunately left unharnessed in practice due to TaqFS and its poor incorporation rate with dITP. Approximately 60% of the cycling time in a BigDye reaction is the incubation time (i.e., 4 min extensions). The fast ramp time afforded with capillary cycling will at most gain a lab 40% in cycling throughput. There are some proposed labor savings with these instruments as they automate the liquid handling and thermal-cycling labor by providing a walk-away system (1800 samples/24 hour).

For close to the same price of $150,000, one can buy a Beckman FX (40 384-well plates sequencing reaction setup/h, 20 384-well SPRI dye terminator purification/h) and four ABI Viper thermal cyclers (24–32 384-wells/day) providing a combined throughput that far exceeds the current Parallab 350 at 50-fold dilutions.

Considering the FX throughput in relationship to a conventional thermal cycler throughput (like the ABI viper), obvious throughput synchronization becomes very evident. One or two FXs could drive a whole a genome center's reaction assembly needs, while 80 cyclers would be required to keep pace. When operations are this poorly impedanced-matched, it rarely makes economic sense for a genome center to fully integrate these three steps. Nevertheless, the Parallab system may offer beneficial features for smaller genome centers where walk away, fully integrated systems are more valuable due to the lack of specialized and highly trained personnel.

One also has to consider the cost of cycling equipment. Currently, ABI 384-well cyclers cost $3500 per 384, or $10/well. The average lab easily obtains three runs per day from a cycler and runs them 300 days/year, placing the cost per well to less than 0.3 cents per read, assuming a three-year lifespan of the equipment. In light of the cost drivers mentioned in Table 13.1, it is suggested that these capillary array technologies will have limited impact on the cost per read unless they can drop BigDye costs to the 200-fold dilution level. Few genome centers have reached this dilution level in production environments, but poster presentations show promise of these dilutions being implemented in production in the near future (Tony West, personal communication).

Table 13.1. Current sequence costs.

Item	2003 $	2006 $	2003 Itemized Cost	2006 Projected Itemized Cost
BigDye	0.04	0.01	1,600,000	400,000
DNA prep	0.015	0.005	600,000	200,000
Array amortization	0.02	0.01	800,000	400,000
Polymer/buffer	0.015	0.01	600,000	400,000
Machine amortization	0.15	0.1	6,000,000	4,000,000
Labor	0.05	0.02	2,000,000	800,000
Overhead	0.464	0.248	18,560,000	9,920,000
Total	0.754	0.403	30,160,000	8,060,000

Items are defined as follows: BigDye, assuming a 50-fold dilution of the recommended 8 µL reaction ($2/50 = $.04), but the price per reaction can range up to $6.75 per reaction based on a user's volume; plasmid preparation, either SPRI or templiphi can be stretched to under 2 cents; array amortization, assuming 2000 runs/array; polymer and buffer, these are traditionally not diluted, but the buffer is likely to be home brewed; machine amortization, assuming 3 years and a $350,000 price at 24 runs/day; labor, includes cost of labor and robot amortization per preparation.

Other Means for 200-fold Dilutions

Two strategies to reduce the volume of dye-terminator mix are commonplace:

1. Dilute the dye in such a format that existing liquid handling instruments can still set up the reactions (3 to 15 µL).
2. Reduce reagent volume to nanoliters such that true miniaturization of the reaction occurs thus preserving the reaction kinetics.

The second strategy offers kinetic advantages, as the fluorescent nucleotides and enzyme under-perform when diluted (http://www.genome.ou.edu/proto.html). When the fluorescent nucleotides are diluted the signal drops accordingly, resulting in suboptimal data quality. Furthermore, as the enzyme is diluted, the tolerance for template concentration in the reaction narrows (Jan Kieleczawa, personal communication). To overcome this limitation, some researchers (Agencourt Biosciences; unpublished results) have added less expensive native Taq to the reaction to assist the enzyme incorporating the nucleotide 5'-triphosphates (dNTPs) at low dilution. Likewise, more dNTPs can be added to assist in the dNTP:ddNTP balance of the reaction as it is

Chapter 13. Future of DNA Sequencing

diluted but, unfortunately, being proprietary, the fluorescent nucleotides are reduced according to the dilution level.

There are at least two physical barriers to volume reduction.

1. Most commercially available 384-pipettors have 5% coefficient of variation (C.V.) dispensing 1 µL and 30% C.V. dispensing 500 nL.
2. Most 384-well thermal cycling plates have a well volume of 40 µL, making cycling at 1 µL very prone to evaporation and condensation. It has been demonstrated that this can be detrimental to the reaction kinetics (2).

There have been several proposed solutions to the above problems:

1. The development of nanopipettors.

More information can be found under the following Web sites:

> http://www.brooks.com/pages/288_parallab.cfm
> http://www.innovadyne.com
> http://www.nanoliter.com
> http://www.allegro-technologies.com

Several other companies offer nanopipetting technologies based on capillary action, microsolonoids, acoustic manipulation, and piezo electric dispensing. Very few have successfully been implemented into a high-throughput production environment as of 2004.

2. Mineral oil capping of reactions.

This is an old solution to a modern problem. Conventionally problematic for automated pipetting, recent adaptation of solid phase purification of cycle sequencing reactions has allowed oil-covered reactions to be purified of the oil after cycling (K. McKernan; oral presentation at the Genome Sequencing and Analysis Conference (GSAC), Boston, MA; 2002).

3. Volume reduction lids.

Silicone seals designed to remove the 35 µL of the volume of the 384-well cycling plate. These lids allow scaling down reaction volumes to 2 to 3 µL, which is compatible with existing liquid handling and cycling equipment, effectively leading to 200-fold dilutions of dye-terminator mix.

4. Geometric modification to the cycling plate well.

By molding the well of the thermal cycling plate to curve away from the thermal cycling block, one can create a thermal gradient in the plate that acts like a thermal lid or condensation zone, which confines the reaction to smaller volume (17).

Multiplexing

Because the overhead associated with managing 20 million purifications and 40 million sequencing reactions is more significant than any given reagent, multiplexing reactions becomes the most leveraged approach in which to reduce cost.

Co-transformation or duplex purifications are potential avenues to be pursued in the coming years. Likewise, duplex sequencing, where forward and reverse reads are sequenced in the same thermal cycling well and are subsequently purified and separated, also has been proposed (Lambert; AMS Feb 2nd, 01 Sanger Centre, Agencourt NHGRI 2003 grant application).

Knowing $5 million human genomes will not fulfill scientific demand for the sequence information; thus, one needs to consider alternate technologies in light of having a reference human genome sequence in hand (12, 25). Resequencing the human genome is vastly easier than *de novo* sequencing. As a result, read length is not of ultimate concern. With 17 bp paired end reads one can readily resequence over 75% of the human genome. Read length longer than 25 bp (allowing for 1 to 2 SNPs or errors) provides diminishing returns until 270 bp (the average length of the 1 million degenerate Alu repeats present in the genome) is surpassed (12). LINE (Long Interspersed Nuclear Elements) and SINE (Short Interspersed Nuclear Elements) also offer challenges, but this can be overcome with proper jumping library construction techniques, which will allow paired ends to be sequenced up to 25 bp from 5, 10, 20, 40, and 100 kb fragments (Agencourt Bioscience Corporation; Advances in Genome Biology and Technology, 2004). Utilizing typeIIS restriction enzymes on 40 kb sheared fragments, Malek et al. describe a method called paired-genome sequence tagging or P-GST, where a single sequencing read can obtain forward and reverse mate pair information with just 60 bp of sequence. These paired tags can be concatermized much like SAGE tags for conventional sequencing or feed directly into massively parallel sequencing by synthesis techniques.

Other considerations include the requirement of homopolymer resolution. Conventional Sanger sequencing (22) resolves these up to 20 bp (13; unpublished results, Agencourt Bioscience Corporation), while other sequencing by synthesis techniques currently do not. Some effort is being placed on reversible terminators to improve the homopolymer resolution with sequencing by synthesis (Genovoxx, GbR, Germany; available at: http://www.genovoxx.com). For many applications, like gene expression, karyotyping, genotyping, and re-sequencing, it is unclear how much impact this will have. *De novo* sequencing certainly requires homopolymer resolution for complete and accurate assembly.

Chapter 13. Future of DNA Sequencing

New sequencing technologies can be broadly categorized into two groups: amplification based or single molecule detection methods.

Amplification Based Methods

Parallel Pyrosequencing

An example of moderately ambitious technology (aiming for a 10-fold reduction in sequencing costs) maturing more rapidly than the $1000 genome projects is 454 Corporation's PicoTiterPlate (14). A one million-well picotiter plate (PTP) allows parallel 50 to 100 bp reads to be detected in a flow cell utilizing pyrosequencing. Currently, the instrument is achieving about 10,000 usable reads per run. Although this is an admirable achievement, there are several drawbacks to this approach.

Because pyrosequencing is chemiluminescent, it requires real time detection of the PTP flow cell. Light generated from the pyrosequencing reaction fades quickly and accurate quantitation of the lights amplitude is important to capture and assess multiple base incorporations from homopolymers. As a result, the scalability of the system beyond one million wells can be accomplished in three ways:

1. Increasing the size of the charge coupled device (CCD) array to cover a larger PTP.
2. Decreasing the size of the beads and the size of the PTP wells from 24-µm diameter beads to 2-µm beads.
3. Utilizing a more directed bead packing protocol than the current random bead packing.

The first option requires the larger CCD array to have pixels arrayed in the geometry of a flow cell. This custom CCD array may add significant cost. The second approach presents significant decay in signal intensities. A 24-µm diameter particle has 144-fold more surface area than a 2-µm diameter bead. As reported by Ronaghi (21), for every 100 cleaved diphosphates generated, one photon is recovered in the pyrosequencing reaction. It is unclear if pyrosequencing produces enough light to be scaled down to a bead with 144-fold less surface area.

Smaller beads also have a higher likelihood of being disrupted or dislodged during flow cell operation. This effect, known as "blow over," is related to bead weight, and diphosphate diffusion is cause for serious concern as the beads in the PTP are not fixed in position, but are gravitationally positioned into picotiter wells. If beads are too light, turbulent flow can move a bead from one well to another, confounding the data analysis. In addition to fixing the beads in place, the wells of the PTP are also responsible for limiting diffusion of the cleaved diphosphates. During polymerization, as each base is incorporated, a diphosphate is released that becomes the substrate for the pyrosequencing enzymatic cascade responsible for the production of light. Because the diphosphate is liberated from the molecule, the generation of light does not necessarily occur local to the DNA. If turbulent flow exists, the diphosphate substrate and the generated signal will diffuse away from the bead. This signal diffusion may limit the attainable bead density unless an alternate means of managing diffusion is employed.

The third option of randomly packing more beads in per cell is a simple limitation in Poisson probabilities unless a more directed or quasi-random packing approach (4) is employed. Currently, the packing method relies on the picotiter plate having a very uniform well diameter that must be less than twice the size of the bead. As a result, the use of monodispersed beads is a necessity to avoid multiple beads from populating a single well. Attempts to increase the concentration of the bead packing result in more multibead wells if the beads are not uniformly monodispersed.

A potential limitation to scaling of this technology is the required real-time monitoring of the pyrosequencing reaction. This permanently couples the reaction with the detection steps, which in future embodiments may not be ideal. The real-time monitoring also presents signal processing challenges as parts of the flow cell initiate the reaction before others. In a single tube format, the pyrosequencing reaction requires about 4 seconds to complete (21). As a result, when applied to a larger array via a flow cell where reagent addition is not uniform and simultaneous, long integration times must be used to image the pyrosequencing reaction across the entire array. Due to the real-time imaging requirements the array can't be scanned; thus, large CCD arrays will be required that can image with high frequency and high sensitivity. This will ultimately limit the cost-effectiveness of the instrument's scalability.

Another drawback to chemiluminescent detection is that until a four-color luciferase assay is developed, pyrosequencing will always be incorporating A, C, G, and T nucleotides separately. This requires four times the number of reagent additions (polymerase, apyrase, luciferase, etc.) needed to extend a single base. Likewise, the enzymes are always incorporating nucleotides without other nucleotide competition, which increases the enzymes' misincorporation rate (1, 18).

Polonies

First reported by Chetverina et al. (8) as a viral diagnostic assay, polymerase colonies (polony) were developed by Mitra and Church as they

Chapter 13. Future of DNA Sequencing

revised the technique for fluorescent in-situ sequencing by synthesis or FISSEQ. Polonies, as originally published (20), are amplified single molecules embedded in an acrylamide matrix. Since acrylamide allows reagents to freely diffuse to the amplified loci, sequencing by synthesis reagents can be readily deployed to sequence the PCR amplicons. As a result, both clonal polymerase chain reaction (PCR) and clonal sequence can be obtained with this approach. Higher concentrations of matrix (acrylamide) retard diffusion of the DNA strands during PCR, which maintains clonality in the PCR but ultimately limits the density of the original method. Too high of a template density produces polonies that cross contaminate each other, and, thus, clonality is compromised. Unlike other aqueous multiplex PCR assays, since the Taq enzyme is present throughout the entire matrix in the polony technique (assuming that all templates are of equal length), the amount of product generated in each polony is expected to be consistent between each loci because polonies are not in competition for reagents. Nonetheless, early embodiments had 6 to 50 μm feature sizes, but the size of DNA molecule amplified was inversely proportional to the feature size (i.e., large amplicons generated small polonies and small amplicons generated very large polonies) limiting the feature density of the early embodiment.

More recently, templates coupled to 1 μm magnetic beads have been introduced to the polony technology, creating very discrete feature size while uncoupling the feature size from the amplicon size. The beads are immobilized in a semi-solid acrylamide matrix that permits diffusion of reagents into and out of the array. This circumvents the micromanufacturing required for picotiter plates and eliminates bead or diphosphate blow-over common in the flow cell design of the PTP. Another benefit to the system is that the density of the array is a simple function of bead dilution or packing.

As seen in Figure 13.1, 15,000 to 30,000 features (1 μm beads) can be imaged in a 20× objective view of a microscope slide. A 3 × 1 inch microscope slide contains over 1000 images with a 20× objective, thus producing over 30 million features per slide with a signal to noise ratio exceeding 200. Sequence of 50 bp from both ends of the template would produce the 3 billion bases required for a 1× of the human genome. Monolayer packing of 1-μm magnetic particles suggests that close to 1 billion microparticles can be packed onto a single 3 inch × 1 inch slide, bringing a 10× genome per slide within reach: 1 billion × 30 bp per feature = 30 billion bases.

A noteworthy distinction of this approach is the simplicity of detection. All that is required is a microscope coupled to a 1 Mega-pixel CCD camera (Olympus, Melville, NY). Automated 0.25 μm resolution microscope stages are readily available (Prior, Rockland, MA) and a combined system for detection would be approximately the cost of a high end microarray scanner. Scalability of the system is quite simple and can be

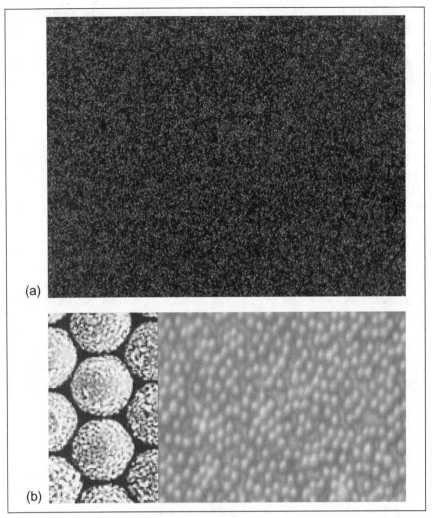

Figure 13.1. Examples of magnetic beads. (a) One-micrometer magnetic beads poulated with an oligo template and a Cy5-incorporated base. Imaged with a 20x objective and a CCD camera. (b) Examples of monolayers of packed 1-μm magnetic beads.

accomplished by either moving to smaller beads while using a 40× objective or simply moving to smaller beads coupled with a higher resolution camera. To date, 20 bp reads have been demonstrated on the system with 99.3% accuracy (personal communication, Jay Shendure).

Critical to accuracy is the type of fluorescent nucleotide utilized. In this respect, Mitra and Church (8) utilized cleavable fluorophores instead of photo-bleachable fluorophores. This is believed to reduce fluorophore quenching and assist the polymerase in nucleotide incorporation. Failure to cleave the dye moiety off the growing strand leaves large bleached fluorophores at each base, which may quench the unbleached incorporated base. These moieties may also sterically hinder the polymerase as it attempts to incorporate additional bases. Cleaving the fluorophore from the base reduces this steric hindrance.

Single Molecule Sequencing by Synthesis

Detection of single molecule incorporation is currently very challenging. Since the pioneering work described by Braslavsky et al. in 2003 (5) for single molecule sequencing, few have since demonstrated more than a few bases of sequence from known templates.

In attempt to eliminate library construction steps, increase density, and increase throughput while reducing reagent consumption, single molecule sequencing by synthesis has been proposed by Solexa (Hinxton, UK) and others (5, 15). In the current non-peer-reviewed Solexa approach, genomic DNA is sheared, size selected, and ligated to linkers. One linker contains a functional group that can be covalently coupled to the glass surface while the other ideally contains a hairpin primer on the free floating 3 prime (3') end. Hairpin primers reduce mispriming of the free 3' end. Without a hairpin primer, the free primer and any internal complementary to the 3' linker will be in competition for single molecule incorporation. Likewise, hairpin primers prevent the primer and extended strand from dissociating from the template during subsequent wash steps.

Care must be taken to properly treat the slides after coupling single-stranded DNA templates, as excess fluorescent nucleotides may adhere to the slide and be detectable as single molecule entities. It is important to emphasize that this technique utilizes single-stranded DNA, so double-stranded read pairing is not possible without significant manipulation of the library ahead of time.

The Solexa system utilizes total internal reflectance (TIR) microscopy. This is needed because a single molecule sequencer requires a very small depth of field to be free of background. The difficulty of this approach lies in the inherent differences in sizes between single molecule (angstroms), laser spot sizes (microns), and the wavelength of fluorescence emission of dyes (nanometers). A scanning laser-PMT would have a spot size that would excite multiple features. A CCD camera could be used, but uniform illumination would require a narrow depth of field to avoid background noise. TIR microscopy achieves a 100-nm sliver of excitation across a 100 µm × 100 µm image. This differs from zero mode waveguides (18) even though the illumination field is less than half the wavelength of the excitation light.

The current configuration according to oral presentations (GSAC, Savannah, 2003), targets 10,000 images per slide at 1-second image acquisition (2.8 hours). Until reversible terminators are fully developed, 100 imaging cycles will be required to obtain an average read length of 43 bases. Because the unknown template randomly shares some sequence with the nucleotide addition sequence, the incorporation rate exceeds one base every four cycles and average one base every 2.3 cycles. For example, if the sequencing by synthesis reagent addition is (GATC) × 25, those templates with GATC stretch (one randomly occurs every 256 bases) incorporates at a rate of one base every incorporation cycle for this sequence stretch. In other words, the worst-case scenario is an incorporation every fourth reagent addition.

Each 100 µm × 100 µm image currently contains 1000 molecules and can be scalable to 10,000. This suggests a molecular density of one molecule every square micron, and with 10 pixels per molecule one assumes a 330 square nm of slide represented by each pixel or approximately the length of 1000 bases. The diameter or coupling footprint of DNA is 20 angstroms, so each pixel is "viewing" 17-fold more empty slide area than the target area.

The highest sensitivity cameras usually sacrifice resolution or utilize pixel binning to achieve their sensitivity. These advertised numbers suggest that a low resolution–high sensitivity CCD camera with a very high objective or magnification is being used in these studies. Working at such magnifications complicates automatic focusing. To process 10,000 images in 2.8 hours, autofocusing needs to be employed, and this becomes much more challenging with the numerical aperture of high magnification objectives.

At the single molecule level fluorescence behaves differently as well. Most fluorophores emit light in a blinking fashion. This is not evident when multiple fluorophores are being illuminated, as this blinking effect appears to be a uniform emission when multiple molecules are blinking stochastically. One also needs to consider how many photons a given fluorophore can emit during the 1-second exposure. Appropriate selection of fluorophores can mitigate the blinking effect but extra attention must be paid to the dyes sensitivity to photobleaching when selecting dyes appropriate for single molecule sequencing.

In addition, it is likely that enzymes will operate much differently in the single molecule environment as the Km (concentration of substrate that produces half-maximal velocity) of an enzyme is usually optimal at micromolar concentrations (15). This requires a high concentration of background nucleotides for efficient and high fidelity enzymatic extension.

Most sequencing by synthesis techniques incorporate one base at a time, since an addition of multiple bases results in the enzyme fully elongating the synthesized strand before adequate detection can occur.

Chapter 13. Future of DNA Sequencing

Single base incorporation allows the enzyme to extend until a different nucleotide is required, thus pausing the polymerization process for ease of detection. This is particularly helpful when an amplification approach is employed, because each copy of a given molecule would be incorporated at a different rate, making real-time detection of the polymerization meaningless. Reversible termination enables all four nucleotides to be present because any given incorporation would terminate all DNA templates in phase.

Since 1994 (18), reversible termination has proven to be challenging. Polymerases have evolved to permit little tolerance for change on the DNA's 3' hydroxyl group. Most strategies for reversible termination have required the addition of cleavable functional groups on the 3' end, thus requiring larger active sites. However, the phenylalanine 667 to tyrosine (F667Y) mutation (23) required for optimal dideoxy sequencing actually reduces the size of the active site of the polymerase. Other approaches (Genovoxx) have utilized large, sterically-hindring cleavable groups linked to the 5' carbon. Once cleaved, the steric hindrance is alleviated and the polymerase can advance. If the steric moiety is a fluorophore, one can accomplish two goals in one step: reversible termination and fluorophore removal. To enable the benefits of reversible termination four distinct fluorophores are needed to infer which nucleotide is incorporated at each cycle.

Levene et al. (15), describe an alternate approach for single molecule DNA detection that offers several signal-to-noise benefits. Conventional single molecule techniques like TIR and fluorescence correlation spectroscopy rely on high spatial resolution to detect single molecule incorporations. Levene et al. (15) demonstrate a technique that results in zeptoliter observation volumes, which enables a high temporal resolution such that real-time polymerization can be observed (Fig. 13.2). Because single molecule detection demands that molecules be packed no closer than the observation volume, zero mode waveguides reduce the observation volume, allowing millimolar enzyme and ligand concentrations to be maintained. The size of the observation volume is directly proportional to the time required for diffusion of reagents in and out of the detection zone. The time required for a fluorescent nucleotide to diffuse into and out of a zeptoliter volume is far less than the time required for an enzyme to incorporate a base. As a result, if a nucleotide is incorporated, the fluorescent nucleotide stalls in the observation volume and emits a detectable signal. By anchoring the fluorophore on the gamma phosphate of the nucleotide, once the enzyme incorporates the base, the fluorophore is liberated and rapidly diffuses from the observation volume. This offers advantages to sequencing by synthesis where each base must be paused for detection and hence requires drastically different nucleotide chemistry.

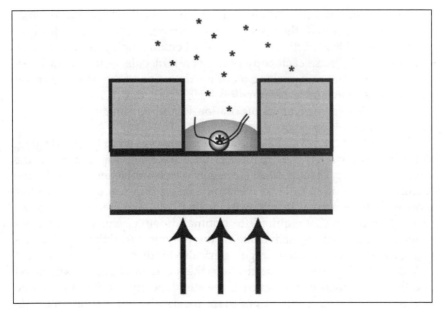

Figure 13.2. The zero-mode waveguide has a high temporal resolution allowing real-time detection of each incorporated base in a zeptoliter volume.

Because the biochemistry of the Levene reaction can function at its desired concentration, the zero-mode waveguide is likely to advance quickly as the biochemistry and enzymology doesn't require drastic recharacterization to function at a nanomolar scale. Nanofluidics, Inc. (Ithaca, NY) is the company focused on commercializing the zero-mode waveguide. To date they have demonstrated single nucleotide sequencing of M13 templates with 60 photons collected per 10-ms bin. The application of phi29 polymerase (3) and gamma-phosphate-labeled nucleotides promises very long reads. To date, few data exist demonstrating the accuracy in sequencing of all four bases. Nonetheless, this is a very promising technology in its early stages of development, which should be monitored closely over the next few years.

Bravslavsky et al. (5) describe an alternate approach to single molecule sequencing. In this approach, a glass slide is pretreated to reduce its affinity for unincorporated nucleotides and to enhance covalent attachment of streptavidin. Biotinylated single-stranded templates are attached to the slides, and a Cy3 oligo is hybridized and imaged. With a 1000× TIR microscope a single molecule can be seen as approximately 1-µm features (Fig. 13.3). To minimize the background of unincorporated nucleotides, single-phase fluorescent resonance energy transfer (spFRET) is utilized.

Chapter 13. Future of DNA Sequencing

Figure 13.3. Single molecule fluorescence. Feature sizes are estimated at 1 to 2 μm based on a published key.

Because the laser only excites the donor molecule on the DNA, only acceptor dyes incorporated in proximity of the donor dye are collected. This ensures that background fluorescence from the laser excitation of the unincorporated acceptor dye is not generated. Although a very impressive accomplishment, the feature size of the image is a bit disappointing. A single DNA oligonucleotide attached to a slide should have diameter of 20 angstroms, and the 1000-base-long DNA fragment would have a length of 340 nm, which is far less than the extrapolated 1- to 2-μm feature size in Figure 13.3. This suggests that pixel binning is being utilized to gain sensitivity that ultimately will limit the spatial density of molecules on the slide or demand very large CCD arrays. In comparison to amplified techniques like polonies, this technique currently offers few density improvements. Single molecule reactions would utilize far fewer reagents, but there are still a few hurdles to overcome before this technique can be applicable. One such drawback is the requirement of FRET and the lack of cleavable dyes. As the technique currently stands, fluorophores are bleached and consecutively labeled incorporations of fluorophores results in marginalized data presumably due to quenching. The development of cleavable fluorophores may overcome this problem. In addition, the FRET has some spatial constraints (usually 80 angstroms) and if it is truly needed for single molecule detection, a donor fluorophore must be incorporated periodically to maintain proximity to the incorporation event. Alternatively, conjugation of the donor dye to polymerase may solve this problem (7, 11). As a result, the current state-of-the-art does not describe DNA sequencing, but DNA fingerprinting of five alternating bases.

Amplification Versus Single Molecule Detection

Amplification-based approaches require far simpler and cheaper detection instrumentation. This significantly reduces signal to noise in comparison to single molecule approaches. The cost of amplification is negligible in light of the $1000 genome goal, and considering the age of the Taq enzyme and PCR patents one should not be shy of leveraging these powerful tools.

Amplification is not free of its own complications. One must consider how to generate clonal PCR products on an addressable array or solid phase. So far, polonies (clonal PCR in an acrylamide matrix) (20) and emulsion or micelle PCR (9) are the only applicable clonal PCR methods described. Emulsion PCR, invented in 1992, is an excellent method for producing billions of microreactors, but it suffers from Poisson distribution problems where some water-in-oil droplets receive two or more template molecules while many droplets receive none. In many respects, an emulsion PCR is very analogous to transformation of E. coli; only there is no selection for micelles with no DNA, emulsion PCR exhibits little to no cloning bias, and E. coli are monodispersed in size. If an emulsion varies in droplet size these Poisson problems become even more complicated. When the emulsion droplet size is too small, PCR is starved of the appropriate reagents to complete PCR. If the emulsion is too large, the likelihood of two templates being present in a droplet is far higher. As a result, generating uniform size emulsions is critical for this technique. Emulsions can be formed that contain magnetic beads. Once an emulsion is formed, and products are amplified and attached to the magnetic bead or solid phase, one needs to harvest the clonal PCR products, which is much cleaner if the PCR products are amplified to one magnetic bead. This requires one to break the emulsion or facilitate the solubility of oil in water. This step is inefficient and some loss of the product is expected. Because emulsion-based microreactors do not allow template strands to diffuse into another microreactor, the density of microreactors can be much higher than the original polony-based approaches where diffusion is only retarded. More importantly, polonies are less likely to suffer from reagent depletion during the PCR than polydispersed emulsions. This has particular dynamic range advantages when selecting a detection schema. Clonal PCR techniques that generate variable product yield, such as nonuniform emulsions, place much more burden on the detection and base calling algorithms required for analysis.

Amplification does present one potential problem for sequencing by synthesis. Because a read out of a base in an amplification-based sequencing by synthesis reaction is an averaged signal from millions of templates all incorporating one base, each incorporation needs to go to >99% com-

Chapter 13. Future of DNA Sequencing

pletion. In other words, all one million strands must go to completion during that extension or degenerate templates would begin to accumulate over the course of the synthesis. Failure to bring to synthesis reaction to completion would create 0.99^{40} template strands sequencing out of phase (n-1 or n-2) after 40 incorporations. This failure to keep the reaction in phase is considered a drawback to most proponents of single molecule sequencing and a key advantage to single molecule sequencers, because single molecule sequencers by definition are asynchronous and cannot get out of phase. Thus, single molecule sequencers do not have to concern themselves with driving each reaction to over 99% incorporation efficiency (usually an asymptotic time curve). As a result, unincorporated strands can simply be picked up on the next appropriate base cycle. Novel sequencers utilizing amplified templates are not able to cut this corner, and longer incorporation steps are expected from these strategies.

However, the single molecules sequencers are not immune to reaction errors. In fact, when errors are made in single molecule systems the sequence signal is indistinguishable from perfect sequence until assembly. If an enzyme has an error rate of 99.999% and there are one billion single molecules on the slide, one has 1000 indistinguishable errors in every incorporation cycle of a plurality of templates. A more significant source of error is likely to be from nucleotide purity and reaction kinetics, which bring the incorporation efficiency in sequencing by synthesis as low as 97%. After 40 incorporations, 70% of the templates on a single molecule slide are either wrong or terminating if one employs a capping method in the sequencing by synthesis. With amplified products the error rates are no better; however, the affected loci manifest themselves with a detectable phenotype incorporating in an n-1 manner and emitting signal in more incorporation reactions than the expected every fourth base. Because the phenotype is predictable, computational subtraction likely mitigates this effect with the amplified approaches. In contrast, single molecule methods produce binary output with limited quality information associated with the base. Four-color reversible terminators enable amplified approaches to measure the signal in all four channels and each incorporation assesses n-1 accumulation and base quality. With single molecule techniques, the assembler has little data from which to differentiate an SNP from an error and is forced to rely on high coverage to cluster real haplotypes from random error in the process. Because enzymes tend to have reproducible misincorporation bias, these errors are not expected to be entirely random. This complicates the ability to utilize coverage to segregate SNPs from misincorporations, as the errors may prove to be reproducible with more coverage. This is particularly important with fluorescent nucleotide purity because the contaminating unlabeled nucleotides can produce deletions with a single molecule sequencer, because certain enzymes have nucleotide specific incorporation bias for unlabeled versus labeled nucleotides.

Another inherent benefit of amplification-based methods is that each incorporation generates more confidence as the detection is very analogous to coincident counting. Because millions of molecules are confirming the event across multiple pixels, it can be certain that the event is real and above noise.

Lastly, amplification-based approaches generate unmethylated templates for sequencing. Single molecule methods are attempting to sequence DNA directly from cells that have methyl cytosines present. Inappropriate fluorescent exposure of this DNA produces C to T transitions, while methyl groups likely complicate the enzyme incorporation kinetics and sequence context-dependent artifacts (6).

Amplification-based approaches are expected to mature more rapidly than single molecule techniques because the detection equipment and the biochemistry are available today. Amplified approaches also offer double-stranded templates more readily and, hopefully, paired end sequencing. Even though single molecule techniques would encompass the ideal sequencing efficiency in respect to reagent consumption, one has to consider whether the redundancy required to mitigate the inherently exclusive single molecule error modes (sensitivity to nucleotide purity, lack of coincident counting, signal-to-noise over unincorporated fluorophores) will effectively force such high coverage upon these systems that they in effect utilize as much reagent and template as amplification-based approaches.

Summary

Sanger sequencing and capillary electrophoresis are not likely to disappear any time soon. The above-described techniques are likely to pioneer new applications and new markets. Since Sanger sequencing has long read lengths and very high fidelity, it will remain the most practical tool for sequence validating a target region of interest. The techniques described in this chapter are primarily focused on resequencing an entire genome and would not be practical to verify a vector, sequence putative kinase mutations in clinical biopsies, or verify a mutagenesis reaction. For this reason, I believe ABI's technology will be around for the long haul. As $100,000 re-sequencing of the genome matures, this is likely to generate new markets as opposed to replace existing ones. In respect to genome centers, it is expected that future genomes will be experimentally drafted with a few-fold coverage of various exploratory technologies. If successful, Sanger sequencing is likely to be maintained for at least a one- to

Chapter 13. Future of DNA Sequencing

threefold coverage of a genome as a high-fidelity scaffold upon which to tile the lower cost, lower fidelity techniques described above.

References

1. Beard, W., Shock, D., and Vande Berg, B. 2002. Efficiency of correct nucleotide insertion governs DNA polymerase fidelity. *J Biol Chem* 277: 47393–47398.
2. Benn, J., and Smith, D. 2003. *U.S. Patent Application 20040009583.* Feb.
3. Blanco, L, Bernad, A., and Salas, M. *U.S. Patent #5,001,050.* March 24, 1989; #5,198,543, March 13, 1991; and #5,576,204, February 11, 1993.
4. Brenner, S., Johnson, M., and Bridgham, J. 2000. Gene expression analysis by massively parallel signature sequencing (MPSS) on microbead arrays. *Nature Biotech* 18: 630–634.
5. Braslavsky, I., Herbert, B., Kartalov, E., et al. 2003. Sequence information can be obtained from single DNA molecules. *Proc Natl Acad Sci USA* 100: 3960–3964.
6. Cantor, C., and Smith, S. 1999. *Genomics: The Science and Technology Behind the Human Genome Project.* New York: John Wiley & Sons Inc.
7. Chan, E. U.S. Patent #6,355,420, August 13, 1998; and #6,263,286, August 13, 1999.
8. Chetverina, H.V., Chetverin, A.B., et al. 1993, Cloning RNA molecules in vitro. *Nucleic Acids Res* 21: 2349–2353.
9. Dressman, D., Yan, H., Traverso, G., et al. 2003. Transforming single DNA molecules into fluorescent magnetic particles for detection and enumeration of genetic variations. *Proc Natl Acad Sci USA* 100: 8817–8822.
10. Ewing, B., Hillier, L., Wendl, M.C., and Green, P. 1998. Base-calling of automated sequencer traces using Phred. II. Error probabilities. *Genome Res.* 8: 186–194.
11. Hardin, S., Gao, X., Briggs, J., et al. 2001. *U.S. Patent Application 20030134807.* December.
12. International Human Genome Sequencing Consortium. 2001. Initial sequencing and analysis of the human genome. *Nature* 409: 860–921.
13. Langan, J.E., Rowbottom, L., Liloglou, T., et al. 2002. Sequencing of difficult templates containing poly A/T tracts, closure of sequencing gaps. *BioTechniques* 33: 276–280.
14. Leamon, J., Lee, W., Tartaro, K., et al. 2003. A massively parallel PicoTiterPlate based platform for discrete picoliter-scale polymerase chain reaction. *Electrophoresis* 24: 3769–3777.
15. Levene, M.J., Korlach, J., Turner, S.W., et al. 2003. Zero-mode waveguides for single-molecule analysis at high concentrations. *Science* 299: 682–686.
16. Malek, J., McEwan, P., Smith, D., et al. 2004. A method for generating paired genome sequence tags. *Advances in Genome Biology and Technology.*
17. Marziali, A. 2001. Patent Cooperation Treaty (PCT) publication number WO 03/006162 A2. July 13.

18. Metzker, M.L., Raghavavhari, R., Richards, S., et al. 1994. Termination of DNA synthesis by novel 3′-modified-deoxyribonucleosides 5′-triphosphate. *Nucleic Acids Res* 22: 4259–4267.
19. Minnick, D.T., Liu, L., Grindley, N.D., et al. 2002. Discrimination against purine-pyrimidine mispaires in the polymerase active site of DNA pol I: a structural explanation. *Proc Natl Acad Sci USA* 99: 1194–1199.
20. Mitra, R.D., and Church, G.M. 1999. In situ localized amplification and contact replication of many individual DNA molecules. *Nucleic Acid Res* 27: 1–6.
21. Ronaghi, M. 2001. Pyrosequencing sheds light on DNA sequencing. *Genome Res* 11: 3–11.
22. Sanger, F., Nicklen, S., and Coulson, A.R. 1977. DNA sequencing with chain-terminating inhibitors. *Proc Natl Acad Sci USA* 74: 5463–5467.
23. Tabor, S., and Richardson, C.C. 1995. A single residue in DNA polymerases of the *Escherichia coli* DNA polymerase I family is critical for distinguishing between deoxy- and dideoxyribonucleotides. *Proc Natl Acad Sci USA* 92: 6339–6343.
24. Tabuchi, M., Masanori, V., and Noritada, K. 2004. Nanospheres for DNA separation chips. *Nature Biotech* 18: 337–340.
25. Venter, J.C., Adams, M.D., Myers, E.W., et al. 2001. The sequence of human genome. *Science* 291: 1304–1351.

Index

ABI. *See* Applied Biosystems
acceptor vectors, 77–78
acetamide, 15–17
acetonitrile, 15–17
Agencourt Bioscience Corporation
    SPRI technology, 101
alcohols, as reagents, 15–17
alkaline lysis procedure, 99–100
Alu repeats, 182
    difficult template sequencing and,
        28–34
Amersham Biosciences
    scanners, 123, 124
    TempliPhi, 101, 104–109
amino acid substitutions, 42
ammonia, as reagent, 16, 17
ampicillin
    culture growth and, 121
    selection, 61
amplification based methods, 183–187,
        192–194
    parallel pyrosequencing, 183–184
    polonies, 184–187
AmpliTaq DNA Polymerase CS,
        37–38
AmpliTaq DNA Polymerase FS, 39
    difficult region reading and, 47
    peak height and, 43, 44–45
animal testing, 159
Applied Biosystems, 3, 124, 136, 137,
        178, 179, 194
    sample sheets and, 136
    scanners, 44, 123, 124–128, 136, 172,
        178
    technology improvement at, 178
archiving data, 172–173
AT-rich DNA
    instability of, 62–63
    pUC vectors and, 56
audits, 174–175
automated DNA scanners. *See*
        scanners, automated DNA
    equipment qualifications for,
        169–170
    overview of, 125, 126–127
    selecting, 125, 128–129

bacteriophage P1 Cre recombinase
        approach
    Creator system, 82–85
    Echo system, 79, 80–82
Baylor University, 79
Beckman CEQ2000, 136
Beckman FX, 179
BigDye Terminator
    benefits of, 39
    Cycle Sequencing Ready Reaction
        Kit, 2
    data quality with, 39–40
    difficult templates and, 32
    dilutions limits of, 13–14, 15, 179
    GC-rich regions and, 28
    peak height and, 43, 44–45
    in plasmid heat-denaturation, 2
    sequence costs of, 180
    v3.0, 2, 11, 12, 15, 29, 32, 39, 45, 46,
        48, 49
billing management, 137

Index

blanks, 166–167
blue/white screens, 56–57, 58, 59–60
Brandis, John W., 35–54
Braslavsky, I., 190–191

calibration, 169
capillary electrophoresis (CE) runs, 146, 151–153
capillary thermal cycling, 178–179
cDNA sequencing, 40
chain-termination method, enzymes in, 35–54
change control, 172
chemiluminescent detection, 183–184
chromatogram files, 136–137
Church, G.M., 184–185, 186–187
Clontech, 79
Code of Federal Regulations, 175
contamination tests, 164
corrective action plans, 165–166
costs
  current, 177
  of plasmid preparation, 112–113
  technology improvement and, 177–196
Creator recombination-based cloning system, 79, 82–85
Cre-loxP recombination system, 79
customer service, 164–167
cycling equipment, 179, 181

data collection and archiving, 172–173
data management systems, 131–142
  billing in, 137
  chromatogram file management in, 136–137
  for core facilities, 133–137
  database updating in, 140
  instrument-related data and, 136
  for large-scale sequencing, 137–141
  over the Internet, 141
  quality control and, 136–137
  sample sheet preparation and, 136
  sample tracking, 133
  security in, 136
  sequence assembly in, 138–140
  sequencing instrument performance and, 137
  sequencing request management, 133
data quality, 35–36
deep-well plates. *See* 96-deep-well plates
de novo sequencing, 182
dGTP dye terminator, 28, 32
dideoxynucleotides
  AmpliTaq DNA Polymerase CS and, 37–38
  enzymes and, 39–40
  labeled, 36–37
  Taq Pol I and, 37–38
difficult templates, sequencing, 27–34
  Alu repeats in, 27–34
  characteristics of, 30
  experimental results in, 31–32
  GC-rich, 27, 28
  homopolymer regions in, 28–34
  materials in, 29
  nucleotide repeats in, 27–34
  protocols for, 29, 31
  strong hairpins in, 28–34
dimethyl sulfoxide (DMSO)
  GC-rich regions and, 28, 32
  as reagent, 15–17
dinucleotide repeats, 58–59
dITP, 39–40, 47
DNA Sequencing Database, 143–156
  analytical operations supported by, 144
  effort reporting in, 154–155
  oligo design in, 147–149
  oligo sequences in, 146–147
  Plate by Column format in, 146, 151–153
  Plate by Row format in, 146, 151–153
  primer inventory in, 147–149
  primer order generation and completion in, 149–150
  Request forms in, 145, 146
  requesting in, 145–147
  results reporting in, 153
  Tube format in, 145–146, 151–153
  workflow models in, 144

Index

documentation, 173–174
donor vectors, 77
d-rhodamine dyes, 38–39
  peak height and, 45
ds DNA, heat-denaturation of, 1–10
dye-labeled terminators, 36–37
  dilution of, 178–181
  enzymes and, 42, 43, 44
  fluorescein, 38–39
  interaction sites and, 42
  rhodamine, 38–39
  searches for improved, 38–39

E. coli, 42
  Pol I, 36
  Pol II, 36
  Pol III, 36
  repeats in, 58–59
Echo recombination-based cloning system, 79, 80–82
EDAS120 Kodak 1D Image Analysis Software, 2
EDTA. *See* ethylenediaminetetraacetic acid (EDTA)
electronic records, 173–174
Elixir of Sulfanilamide, 159
Ellidge, Steve, 79
enzymes, new DNA sequencing, 35–54
  AmpliTaq DNA Polymerase FS and, 37–38
  conclusions regarding, 48–50
  difficult region reading and, 46–48
  dye-labeled terminators and, 36–37, 38–39
  dye primers and, 37–38
  from eubacterial genera, 41
  methods and materials, 43–44
  modified *Taq* Pol I, 41–43
  from noneubacterial genera, 40–41
  peak height and, 43, 44–45
  polymerases, 36
  polymerases, sources of new, 40–43
  read length and, 43, 44
  salt tolerance and, 45–46, 47
  searches for better, 39–40
  sequencing reactions and, 44
equipment qualifications, 169–170
ethanol precipitation, 38

ethylenediaminetetraacetic acid (EDTA)
  in plasmid heat-denaturation, 2, 5
  plasmid storage in, 4
  as reagent, 12–13, 14

facility requirements, 167–168
false positives/negatives, 59–60
Federal Food, Drug and Cosmetic Act (1938), 159
Finch-Server, 131–142
  Assembly Manager, 132, 139
  benefits of, 131–132
  billing and, 137
  BLAST Manager, 132, 139, 141
  Chromatogram Manager, 136–137
  components of, 132
  Core DNA Sequencing System, 132, 134–135
  database updating and, 140
  Internet data management with, 141
  quality control and, 136–137
  sample sheets and, 136
  sample tracking with, 133
  security and, 136
  sequence assembly and, 138–140
  sequencing instrument performance and, 137
  Sequencing Request Manager, 132, 133
  tables in, 132–133
fluoresceins, 38–39
fluorescent *in-situ* sequencing by synthesis (FISSEQ), 184–185
fluorophores, 186–187
Food and Drug Administration. *See* U.S. Food and Drug Administration
foreign DNA/RNA
  reagent reactions and, 17–18
4D, Inc., 143

Gateway recombination-based cloning system, 79, 85–94
GC-melt reagent, 32
GC-rich regions, 27–34
Gene Codes Corporation, 153
genes of interest (GOIs), 77, 78

Index

Geospiza Inc., Finch-Server, 131–142
glycerol, 15–17
Godiska, Ronald, 55–75
Godlevski, Michele, 157–176
good laboratory practices (GLP), 157–176
　applicability of, 160
　change control in, 172
　corrective and preventative action in, 165–167
　customer service and, 164–167
　data collection/archiving in, 172–173
　documentation in, 173–174
　electronic records in, 173–174
　equipment qualifications in, 169–170
　facility requirements in, 167–168
　history of, 158–160
　implementation issues of, 175
　information resources on, 175–176
　inspections/audits in, 174–175
　maintenance and calibration in, 169
　personnel in, 168
　process/system validation in, 170–171
　quality assurance and, 162–164
　reagents and solutions in, 168–169
　standard operating procedures and, 160–162
　training in, 168
good manufacturing procedures (GMP), 157–176
　history of, 158–160
GXP regulations, 167–168

hairpins, difficult template sequencing and, 28–34
heat-denaturation of plasmids, 1–10
　benefits of, 6–9
　buffer strength-dependence of, 4
　DNA sequencing in, 2–3
　experimental design for, 2
　foreign DNA/RNA and, 17–18
　plasmid size in, 4–6
　temperature-dependence of, 3
　time course of, 4–6
Herbert, B., 190–191

homopolymer regions
　difficult template sequencing and, 28–34
　enzymes and, 40
　resolution of, 182
Human Genome Project, 27
hyperthermophilic polymerases, 40–41

iFinch, 141
InFusion recombination-based cloning system, 79, 83, 85, 95, 97
injection control, 164
insert-driven transcription, 60
inspections, 174–175
installation qualification (IQ), 170
interaction sites, 42
Internet, data management via, 141

Jungle, The (Sinclair), 158

Kalman, L.V., 38
Kartalov, E., 190–191
Kefauver-Harris Amendments (1963), 159
Kennedy, Edward, 159
Kieleczawa, Jan, 1–34, 123–129
Koffman, Donald M., 143–156
Korlach, J., 189–190

laboratory information management systems (LIMS), 173
$lacZ\alpha$, false positives/negatives and, 59–60
lambda-phage recombinase approach, 79, 85–94
lane markers, 163
lane tracking lines, 172–173
Levene, M.J., 189–190
Life Technologies, 79
linear DNA amplification, 101, 104
Ling, Vincent, 77–98
Long Interspersed Nuclear Elements (LINE), 182

magnesium ($MgCl_2$)
　BigDye dilution and, 14
　as reagent, 12, 13
　maintenance issues, 169
McKernan, Kevin, 177–196

Index

MCPrep, 101
Mead, David A., 55–75
MegaBACE scanners, 124–125, 136
   sequencing results with TempliPhi, 110–111
microscaling, 178–179
mineral oil capping, 181
Mitra, R.D., 184–185, 186–187
multiplexing purifications, 182

nanopipettors, 181
nanoscaling, 178–179
NaOH-induced plasmid denaturation, 1–10
96-deep-well plates, 101, 117–122
   antibiotic amount and, 121
   glass tubes vs., 121–122
   growth culture volume and, 119, 121
   growth length and, 119
   growth media and, 118–119
   shaking speed and, 118–119
noneubacterial genera, polymerases from, 40–41
Nonident P-40 detergent, difficult templates and, 28

O-helix, 42
oligos
   definition of, 146–147
   designing, 147–149
   ordering, 147, 149–150
open reading frames (ORFs), 57–58
operational qualification (OQ), 170

Parallab, 178–179
parallel pyrosequencing, 183–184
patent medicines, 158
Patterson, Melodee, 55–75
pBlueScript vectors, 70
PCR. *See* polymerase chain reaction (PCR)
peak height
   BigDye Terminator v3.0 and, 39–40
   enzymes and, 43, 44–45
   Taq Pol I and, 37–38
performance qualification (PQ), 170
personnel issues, 168

pEZSeq vectors, 67
pGEM vectors, 70
Phrap, 138, 140
Phred, 19, 47, 48, 137
picotiter plates (PTPs), 183
plasmids
   culturing methods for, 109–111
   heat-denaturation of, 1–10
   high-throughput preparation of, 100–101
   low-throughput preparation of, 99–100
   mobilization of, 61–62
   96-deep-well culture of, 117–122
   96-well purification systems, 101
   preparation costs of, 112–113
   preparation methods for, 99–115
   products for preparation of, 102–103
   pUC, 55–75
   TempliPhi and, 101, 104–109
polonies, 184–187
polyA regions, difficult template sequencing and, 28–34
polymerase chain reaction (PCR)
   plasmid heat-denaturation for, 1–10
   polonies and, 185
   Taq Pol I in, 36
polymerases
   archael, 40–41
   classes of, 36
   from eubacterial genera, 41
   modified *Taq* Pol I, 41–43
   from noneubacterial genera, 40–41
   polonies, 184–187
polymerization control, 163
polyT regions, difficult template sequencing and, 28–34
Porter, Sandra, 131–142
positive control samples, 163
primers
   concentration of, and reagent reactions, 18, 19
   dye-labeled terminators vs., 36–37
priming sites, foreign DNA/RNA and, 17–18
process validation, 170–171
pSMART vectors, 65–67, 69
pTrueBlue vectors, 68–70

pTZ vectors, 70
pUC18, 55
pUC vectors, 55–75
  alternatives to, 63–70
  ampicillin selection and, 61
  AT-rich DNA instability and, 62–63
  blue/white screens and, 56–57, 59–60
  drawbacks of, 56–63
  false positives/negatives in, 59–60
  high copy number of, 61
  insert-driven transcription and, 60–61
  open reading frames and, 57–58
  plasmid mobilization and, 61–62
  repeats and, 58–59
  replication and, 59
Pure Food and Drug Act (1906), 158–159
pyrosequencing, 183–187
pZErO vectors, 70

quality control, 136–137, 157–176
  for capillary-based systems, 163–164
  customer service and, 164–167
  for gel-based systems, 163
  good laboratory practices on, 162–164
Quality System Inspection Technique, 175

RCA (rolling circle amplification), 101, 104–109
reaction miniaturization, 178–179
read lengths
  base callers and, 128–129
  data quality in, 35–36
  enzymes and, 43, 44
  in heat-denaturation, 6, 7, 8
  magnesium and, 12
  primer concentration and, 18, 19
  salts and, 14–15, 16
reagents, 11–26
  acetamide, 15–17
  acetonitrile, 15–17
  alcohols, 15–17
  ammonia, 16, 17
  DMSO, 15–17

EDTA, 12–13, 14
experimental design for, 12
foreign DNA/RNA and, 17–18
GC-melt, 32
glycerol, 15–17
good laboratory practices and, 168–169
$MgCl_2$, 12, 13
primer concentration and, 18, 19
salts, 14–15, 16
troubleshooting reactions with, 18, 20–25
recombination-based cloning, 77–98
  bacteriophage P1 Cre recombinase, 79, 80–82
  Creator system of, 79, 82–85
  Echo system of, 79, 80–82
  future of, 94–98
  Gateway system of, 79, 85–94
  InFusion system of, 79
  lambda-phage recombinase, 79, 85–94
  primary considerations in, 77–80
  system selection in, 79–80
repeats
  Alu, 28–34, 182
  difficult template sequencing and, 28–34
  dinucleotide, 58–59
  in *E. coli*, 58–59
  pUC vectors and, 58–59
  trinucleotide, 58–59
replication, origin of, 59
reversible termination, 189
rhodamines, 38–39
  peak height and, 45
Richardson, C.C., 37
rich-growth media, 118–119
RNAi technology, difficult templates and, 28
robustness tests, 171
rolling circle amplification (RCA), 101, 104–109
ruggedness tests, 171

salts
  enzyme tolerance of, 45–46, 47
  as reagents, 14–15, 16
sample sheets, 136

Index

scanners, automated DNA, 123–129, 169–170
Schoenfeld, Tom, 55–75
Searle Laboratories, 159
security issues, 136
sequence assembly, 138–140
Sequencher, 153
shotgun libraries, construction of, 55–75
Sinclair, Upton, 158
single molecule sequencing, 187–191, 193–194
single-phase fluorescent resonance energy transfer (spFRET), 190–191
siRNA vector plasmids, difficult templates and, 28
Slagel, Joe, 131–142
Smith, Todd, 131–142
Solexa system, 187–191
Sookdeo, Hemchand, 143–156
SPRI technology, 101
spSMART C vector, 67–68
Spurgeon, Sandra L., 35–54
SQL (Structured Query Language), 133
ss DNA, from ds DNA, 1–10
standard operating procedures (SOPs), 160–162
system validation, 170–171

Tabor, S., 37
Taq chemistries
  AmpliTaq DNA Polymerase CS, 37, 39
  F667Y, 41
  GC-rich regions and, 28
  G46D, 37
  nuclease structure in, 37–38
  Pol I, 36, 37–38
  Pol I, modified, 41–43
  Pol I, structure of, 49–50
  Tma Pol I, 45, 47, 48–49
  Tth Pol I, 45, 47, 48
Taylor, Thalia, 157–176
T-blob problems, 167
technology
  amplification based, 183–187, 192–194
  capillary thermal cycling, 178–179

future of, 177–196
improvements in existing, 178
multiplexing, 182
parallel pyrosequencing, 183–184
polonies, 184–187
sequencing reaction miniaturization and, 178–179
single molecule sequencing, 187–191, 193–194
TempliPhi, 101, 104–109
  benefits of, 111
  culturing methods compared with, 109–111
  protocols for, 105, 106–107
  reaction products with, 105, 108–109
  terminators. *See* BigDye Terminator; dye-labeled terminators
Thalidomide, 159
Thermotoga species, 41, 42–43
Thermus species, 41, 42–43
throughput, 35
Tma Pol I, 45, 47, 48–49
total internal reflectance (TIR) microscopy, 187–188
training, 168
transcription, insert-driven, 60
trinucleotide repeats, 58–59
troubleshooting reagent reactions, 18, 20–25
T664G, 45
Tth Pol I, 45, 47, 48
Turner, S.W., 189–190

United States Pharmacopoeia, 159
unstable DNA, vectors for cloning, 55–75
  pBlueScript, 70
  pEZSeq, 67
  pGEM, 70
  pSMART, 65–67
  pSMART C, 67–68
  pTrueBlue, 68–70
  pTZ, 70
  pUC, drawbacks of, 56–63
  pZErO, 70
U.S. Food and Drug Administration
  GLP and, 157
  history of, 158–160

inspections/audits, 174–175

volume reduction lids, 181

Wiley, Harvey, 158

Wiley Act of 1906, 158–159

Wyeth Research, Core Development DNA Sequencing Services, 143

zero-mode waveguides, 189–190